AF361850

Transport and Diffusion in Quantum Complex Systems

Transport and Diffusion in Quantum Complex Systems

Editors

Paolo Bordone
Dario Tamascelli

MDPI • Basel • Beijing • Wuhan • Barcelona • Belgrade • Manchester • Tokyo • Cluj • Tianjin

Editors
Paolo Bordone
Università degli Studi di
Modena e Reggio Emilia
Italy

Dario Tamascelli
Università degli Studi di Milano
Italy

Editorial Office
MDPI
St. Alban-Anlage 66
4052 Basel, Switzerland

This is a reprint of articles from the Special Issue published online in the open access journal *Entropy* (ISSN 1099-4300) (available at: http://www.mdpi.com).

For citation purposes, cite each article independently as indicated on the article page online and as indicated below:

LastName, A.A.; LastName, B.B.; LastName, C.C. Article Title. *Journal Name* **Year**, *Volume Number*, Page Range.

ISBN 978-3-0365-2181-7 (Hbk)
ISBN 978-3-0365-2182-4 (PDF)

Contents

About the Editors

Paolo Bordone, Ph.D., is Associate Professor of Theoretical Physics at the Department of Physics, Informatics and Mathematics of the University of Modena and Reggio Emilia. He is author and co-author of more than 90 publications in international journals. His main research interests are: charge quantum transport in nanostructures, electron quantum optics, quantum walks, coherent quantum dynamics and quantum dynamics of noisy systems, quantum correlations in complex systems, quantum estimation theory.

Dario Tamascelli, Ph.D., is Assistant Professor at the Department of Physics of the University of Milan, Italy, and research fellow at the Institute of Theoretical Physics and Center for Integrated Quantum Science and Technology (IQST) of the Ulm University (Germany). His research interests are Quantum walks, Quantum transport, Quantum biology, Characterization and Simulation of Open Quantum Systems, Quantum metrology, GPU-computing, and Tensor Network techniques.

Preface to "Transport and Diffusion in Quantum Complex Systems"

The understanding of transport of energy, mass, charge or information in complex quantum systems plays a key role from both a fundamental and technological point of view. As such, it is triggering a large amount of theoretical and experimental research aimed to understand and exploit quantum coherent phenomena for the development of quantum devices possibly outperforming their classical counterpart.

Quantum interference is at the origin of a number of peculiar effects such as, for example, ballistic transport along lattices and resonant tunneling. On the other side, the presence of unavoidable interactions with surrounding environment typically leads to loss of coherence and to the emergence of a diffusive behavior, closer to the classical scenario, and, in some cases, even enhancing the transport efficiency. The control of such phenomena, together with the understanding of the transition form microscopic to macroscopic or from single-particle to few- or many-particle systems, is of utmost importance for the successful build-out of quantum technologies, and constitutes the focus of this book.

Paolo Bordone, Dario Tamascelli

Editors

Article

Matrix Product State Simulations of Non-Equilibrium Steady States and Transient Heat Flows in the Two-Bath Spin-Boson Model at Finite Temperatures

Angus J. Dunnett * and Alex W. Chin

Institut des NanoSciences de Paris, CNRS, Sorbonne Université, 4 Place Jussieu, 75005 Paris, France; alex.chin@insp.upmc.fr
* Correspondence: angus.dunnett@insp.upmc.fr

Abstract: Simulating the non-perturbative and non-Markovian dynamics of open quantum systems is a very challenging many body problem, due to the need to evolve both the system and its environments on an equal footing. Tensor network and matrix product states (MPS) have emerged as powerful tools for open system models, but the numerical resources required to treat finite-temperature environments grow extremely rapidly and limit their applications. In this study we use time-dependent variational evolution of MPS to explore the striking theory of Tamascelli et al. (Phys. Rev. Lett. **2019**, *123*, 090402.) that shows how finite-temperature open dynamics can be obtained from zero temperature, i.e., pure wave function, simulations. Using this approach, we produce a benchmark dataset for the dynamics of the Ohmic spin-boson model across a wide range of coupling strengths and temperatures, and also present a detailed analysis of the numerical costs of simulating non-equilibrium steady states, such as those emerging from the non-perturbative coupling of a qubit to baths at different temperatures. Despite ever-growing resource requirements, we find that converged non-perturbative results can be obtained, and we discuss a number of recent ideas and numerical techniques that should allow wide application of MPS to complex open quantum systems.

Keywords: open quantum systems; tensor networks; non-equilibrium dynamics

Citation: Dunnett, A.J.; Chin, A.W. Matrix Product State Simulations of Non-Equilibrium Steady States and Transient Heat Flows in the Two-Bath Spin-Boson Model at Finite Temperatures. *Entropy* **2021**, *23*, 77. https://doi.org/10.3390/e23010077

Received: 26 November 2020
Accepted: 25 December 2020
Published: 6 January 2021

Publisher's Note: MDPI stays neutral with regard to jurisdictional claims in published maps and institutional affiliations.

1. Introduction

The physics of open quantum systems (OQS) plays a critical role in almost all aspects of quantum science [1,2], and the emergent phenomena of dephasing, decoherence and dissipation particularly limit our ability to initialise and control multi-partite quantum states. As a direct result of this, the development of scalable quantum technologies is greatly constrained by open system phenomena, and understanding how irreversibility arises from microscopic system-environment interactions has become essential for finding ways to mitigate deleterious noise effects [3]. However, alongside this goal of suppressing dissipative noise—normally by making the systems less 'open'—the theory of OQS also plays a vital role in the design of systems where the exploitation of strong energy and information exchange between a system and its environment is desirable: this is the world of quantum thermodynamics and nanoscale energy harvesting, storage and transduction [4–6].

Any 'machine' or device capable of converting ambient energy into work must necessarily be an open system. As these machines shrink to lengths where such energetic transformations can become few-quanta, ultra-fast events, it becomes necessary to describe their functional dynamics on timescales over which system-environment correlations—in both space and time—may be highly relevant [7,8]. Unlike the perturbative OQS found, for example, in atomic systems where dissipation can be characterised by simple decay rates, quantum energy harvesting naturally focuses on the highly non-Markovian and non-perturbative regime of OQS where the border between the 'system' and 'environment' degrees of freedom is ill-defined. Moreover, as systems capable of converting thermal energy must also reject a certain amount of heat to a colder reservoir [6], the study of quantum energy harvesting leads directly to a consideration of multi-environment OQS, and the extended, inter-environmental quantum correlations that could be generated under non-equilibrium operating conditions.

Molecular and biological light-harvesting systems provide a good example of such nanoscale energy extraction, in which a non-thermal population of electronic excitations (excitons, charge pairs, etc.) appears from the molecule-mediated connection of the 'hot' photon and 'cold' vibrational environments. In this context, much attention has been placed on the complex physics due to the strong coupling and non-separable timescales of electronic and environmental (vibrational) dynamics [9–11], which include potentially exploitable effects such as transient breaking of detailed balance [12], noise-induced electronic coherence and cooperative multi-environment effects [13,14]. In such studies, the effect of light is normally assumed to be weak, leading to the 'additive' approximation that phenomenological terms describing excitation, emission and dephasing can be simply added to the more complex equations of motion of the vibronic open system. However, organic molecules often have very strong light-matter coupling and can show surprising non-additive effects [15,16], including nonlinear polaritonic weakening of exciton-phonon coupling in micro-cavity systems [17].

The example above highlights the theoretical challenges posed by some energy harvesting systems: non-perturbative and highly structured couplings, comparable dynamical timescales and competing environmental processes. Under these conditions the dissipative dynamics of the system's reduced density matrix cannot be simply described by dephasing and relaxation rates: the full real-time evolution of the system and its environments must be accounted for on an essentially equal footing. This looks, a priori, like a hopeless task, as each environment contains a continuum of quantum excitation modes, and the formal number of quantum states in any computation will explode exponentially with the number of such modes. However, things are not so desperate, and two broad responses to this problem have emerged over recent years: one branch aims to efficiently simulate the propagators of the system's reduced density matrix [1,9,18,19], the other aims at representing and evolving the entire system-environment wave function. Important contributions in this latter domain are Density Matrix Renormalization Group (DMRG) techniques such as the Time Evolving Density operator with Orthogonal Polynomials Approach (TEDOPA) [12,20], Dissipation-Assisted Matrix Product Factorization [21], Time-Dependent Numerical Renormalisation Group techniques and the Multi-Layer Multi Configurational Time-Dependent Hartree method (ML-MCTDH) developed in chemical physics [22,23].

The key to all of the wave function methods is the observation that, given a well-defined initial condition, the quantum dynamics generated by typical system-environment Hamiltonians leave the state inside a much smaller sub-space of the complete Hilbert space of the problem. This suggests that the wave function can be parameterised by a potentially tractable number of parameters, and—as we shall see—the effectively short-range, one-dimensional structure of OQS Hamiltonians implies that Matrix Product States (MPS) will provide an efficient and versatile format for many system-environment wave functions. Viewed this way, the parameters (matrices) of an MPS can be considered as variational degrees of freedom, leading to the powerful 1-site time-dependent variational principle (1TDVP) algorithm for efficient propagation of large wave functions in real-time [24]. This general technique can be used in any MPS and Tree-Tensor Network problem [14], but its particular utility in open-system problems has only recently been appreciated. We shall make use of this technique in this article, but a discussion of MPS, TDVP and other computational aspects is left to the dedicated presentations in the literature [24–27].

Instead, the key issue that we wish to explore in this study is the remarkable recent result of Tamascelli et al. [28] that allows wave function approaches to OQS to effectively capture the effects of finite temperature environments through the simulation of an equivalent zero-temperature proxy system. As already discussed above, in the non-perturbative, non-Markovian regime of OQS, computing the evolution of the single wave function from a sharp initial condition can already be very demanding: converging results over the astronomically large space of initial conditions in a thermal ensemble rapidly becomes impossible. If we also wish to explore the role of non-classical effects in heat flows between finite-temperature environments, the problem becomes exponentially worse. The access to finite temperature properties from a single zero-temperature (pure) wave function simulation thus opens up an entire class of powerful non-perturbative methods for the study of novel open quantum systems. This work aims to establish the extent to which Tamascelli's 'T-TEDOPA' theory translates into affordable non-perturbative TDVP simulations of thermal and non-equilibrium OQS dynamics, as well as to explore some of the non-classical and non-additive aspects of heat exchange in OQS.

This article is organised as follows. In Section 2.1 we present the spin-boson Hamiltonians that we will simulate. Sections 2.2 and 2.3 give a summary of the T-TEDOPA theory that we will employ in our numerical investigations. Section 2.4 then presents a careful study of the non-perturbative spin-boson model at finite temperatures which reveals some of the practical numerical costs implicit in this approach. Thanks to this testing, we are able to offer a freely accessible dataset that can be used as a benchmark for other numerical approaches to this model, as well as code packages that allow users to perform their own TDVP calculations on finite-temperature open systems. In anticipation of the need to explore non-equilibrium states in a wide range of future contexts, we go on to test the non-perturbative physics of a two-level system (TLS) coupled to two environments at different temperatures in Section 2.5. Exploiting the information in the many-body system-environment(s) wave function, we examine the microscopic behaviour of the heat flows between the system and the environments as a function of environmental coupling strength and temperature differences, and highlight a number of non-additive effects arising from non-perturbative quantum polaron effects. Finally, we summarise and discuss our findings in Section 3.

2. Results

2.1. Model, Parameters and Initial Conditions

We shall base our exploration of finite temperature open dynamics on numerical simulations and analysis of a quantum two-level system that is strongly coupled to either one or two baths of bosonic harmonic oscillators, as illustrated in Figure 1a. The two baths are labelled a and b and are at different inverse temperatures β_a and β_b, respectively ($\beta = 1/(k_b T)$). The system-bath Hamiltonian is given by

$$\hat{H} = \frac{\omega_0}{2}\sigma_z + \hat{H}_I^a + \hat{H}_I^b + \hat{H}_B^a + \hat{H}_B^b, \tag{1}$$

where

$$\hat{H}_I^a = \sigma_x \otimes \sum_k (g_k^* \hat{a}_k + g_k \hat{a}_k^\dagger) \tag{2}$$

$$\hat{H}_I^b = \sigma_x \otimes \sum_k (g_k^* \hat{b}_k + g_k \hat{b}_k^\dagger) \tag{3}$$

$$\hat{H}_B^a = \sum_k \omega_k \hat{a}_k^\dagger \hat{a}_k \tag{4}$$

$$\hat{H}_B^b = \sum_k \omega_k \hat{b}_k^\dagger \hat{b}_k. \tag{5}$$

The TLS is described by the standard Pauli operators σ, while the $\hat{a}_k (\hat{b}_k)$ are bosonic anihilation operators for harmonic modes of frequency ω_k in bath $a(b)$. The corresponding creation operators are denoted $\hat{a}_k^\dagger (\hat{b}_k^\dagger)$. The k harmonic of each bath couples to the TLS with a coupling strength denoted g_k, which we take to depend on the index k but not on a or b.

The spectral density of the environment is defined as $J(\omega) \equiv \pi \sum_k |g_k|^2 \delta(\omega - \omega_k)$, where $\delta(x)$ is the Heaviside Theta function. As a smooth, continuous function of frequency, the spectral density can take various forms in specific physical realisations such as electron-phonon interactions, emitter-photon or exciton-vibration coupling in molecular systems. It is well known that the qualitative behaviour of the TLS depends sensitively on the form of $J(\omega)$, especially at low temperatures [1]. For simplicity, we assume identical system-bath couplings for both environments and use the common linear frequency dependence that defines an Ohmic environment

$$J(\omega) = 2\pi\alpha\omega\theta(\omega_c - \omega),\tag{6}$$

where α is a dimensionless coupling constant and we have introduced a hard frequency cut-off ω_c.

The initial condition $\hat{\rho}(0)$ for our numerical simulations is taken to be an uncorrelated (product) state of the spin and baths, which—because of the baths' finite temperatures—must be described by a mixed state, i.e., a density matrix

$$\hat{\rho}(0) = \rho_s \otimes \frac{e^{-\hat{H}_B^a \beta^a}}{\mathrm{Tr}\{e^{-\hat{H}_B^a \beta^a}\}} \otimes \frac{e^{-\hat{H}_B^b \beta^b}}{\mathrm{Tr}\{e^{-\hat{H}_B^b \beta^b}\}},\tag{7}$$

where ρ_s is some arbitrary initial density matrix for the TLS.

Remarkably, despite the initial condition containing two statistically mixed thermal density matrices, it has recently been shown by Tamascelli et al. that the reduced dynamics of the spin can still be obtained from a single simulation of an equivalent pure, i.e., zero temperature, system-environment wave function [28,29]. As this result is central for generating our numerical results and our later discussion, we shall now give a brief summary of the protocol first presented in Ref. [28].

2.2. Finite-Temperature Reduced Dynamics from Pure Wave Function Evolution

In this section we shall closely follow the original notation and presentation of Tamascelli et al. [28] and, to simplify the presentation, we shall only consider the coupling to a single environment denoted E. The procedure can be easily generalised to multiple environments. Our starting point is the generic Hamiltonian for a system coupled to a bosonic environment consisting of a continuum of harmonic oscillators

$$H_{SE} = H_S + H_E + H_I,\tag{8}$$

where

$$H_I = A_S \otimes \int_0^\infty d\omega \hat{O}_\omega, \quad H_E = \int_0^\infty d\omega\omega a_\omega^\dagger a_\omega.\tag{9}$$

The Hamiltonian H_S is the free system Hamiltonian and A_S is a generic system operator which couples to the bath. The environment's free Hamiltonian is given by H_E. For the bosonic bath operators we take the displacements

$$O_\omega = \sqrt{J(\omega)}(a_\omega + a_\omega^\dagger),\tag{10}$$

thus defining the spectral density $J(\omega)$. This has been written here as an arbitrary continuous function, but we note that the formulas can also be applied to the case of coupling to a discrete set of vibrational modes by adding suitable structure to the spectral density, i.e., sets of lorentzian peaks or Dirac functions [30–32].

The state of the system+environment at time t is a mixed state described by a density matrix $\rho_{SE}(t)$. The initial condition is assumed to be a product of system and environment states $\rho_{SE}(0) = \rho_S(0) \otimes \rho_E(0)$ where $\rho_S(0)$ is an arbitrary density matrix for the system and $\rho_E(0) = \exp(-H_E\beta)/\mathcal{Z}$, with the environment partition function given by $\mathcal{Z} = \mathrm{Tr}\{\exp(-H_E\beta)\}$. Such a product state is commonly realised in non-equilibrium problems where the system is suddenly prepared or projected into an excited state from a ground state in which the system and environment states are separable. This type of preparation is exemplified by the Franck-Condon principle in molecular photophysics, where the optical transition occurs without any change in the nuclear degrees of freedom, leaving the subsequent relaxation dynamics to evolve from a product 'initial' condition [33,34]. The environment thus begins in a thermal equilibrium state with inverse temperature β, and the energy levels of each harmonic mode are statistically populated. For a very large number (continuum) of modes, the number of possible thermal configurations grows extremely rapidly with temperature, essentially making it impossible to obtain a converged sampling of these configurations when each instance involves demanding wave function simulations. We briefly note that some more efficient sampling methods involving sparse grids and/or stochastic mean-field approaches have recently been proposed and demonstrated [35,36], as well as some effective MPS techniques for capturing finite temperature effects in frequency domain simulations [37].

The initial thermal condition of the environmental oscillators is also a Gaussian state, for which it is further known that the influence functional [1]—which is a full description of the influence of the bath on the system—will depend only on the two-time correlation function of the bath operators

$$S(t) = \int_0^\infty d\omega \langle O_\omega(t) O_\omega(0) \rangle. \tag{11}$$

Any two environments with the same $S(t)$ will have the same influence functional and thus give rise to the same reduced system dynamics, i.e., the same $\rho_S(t) = \text{Tr}\{\rho_{SE}(t)\}$. That the reduced density matrix's dynamics are completely specified by the spectral density and temperature of a Gaussian environment has been known for a long time [1], but the key idea of the equivalence—and thus the possibility of the interchange—of environments with the same correlation functions has only recently been demonstrated by Tamascelli et al. [29].

The time dependence in Equation (11) refers to the interaction picture so that the bath operators evolve under the free bath Hamiltonian: $O_\omega(t) = e^{iH_E t} O_\omega(0) e^{-iH_E t}$. Using Equation (10) and $\langle a_\omega^\dagger a_\omega \rangle = n_\beta(\omega)$ we have

$$S(t) = \int_0^\infty J(\omega)[e^{-i\omega t}(1 + n_\beta(\omega)) + e^{i\omega t} n_\beta(\omega)]. \tag{12}$$

Making use of the relation

$$\frac{1}{2}(1 + \coth(\omega\beta/2)) \equiv \begin{cases} n_\omega(\beta), \omega \geq 0 \\ -(n_{|\omega|}(\beta) + 1), \omega < 0 \end{cases} \tag{13}$$

we can write Equation (12) as an integral over all positive and negative ω

$$S(t) = \int_{-\infty}^\infty d\omega \text{Sign}(\omega)\frac{J(|\omega|)}{2}(1 + \coth(\frac{\omega\beta}{2}))e^{-i\omega t}. \tag{14}$$

However, Equation (14) is exactly the two-time correlation function one would get if the system was coupled to a bath, now containing positive and negative frequencies, at zero temperature! The effects of the finite, physical temperature now appear in a new effective spectral density for the extended environment given by

$$J_\beta(\omega) = \text{Sign}(\omega)\frac{J(|\omega|)}{2}(1 + \coth(\frac{\omega\beta}{2})). \tag{15}$$

Thus, we find that our open system problem is completely equivalent to the one governed by the Hamiltonian

$$H = H_S + H_E^{\text{ext}} + H_I^{\text{ext}}, \tag{16}$$

in which the system couples to an extended environment, where

$$H_I^{\text{ext}} = A_S \otimes \int_{-\infty}^{\infty} d\omega \sqrt{J_\beta(\omega)}(a_\omega + a_\omega^\dagger),$$

$$H_E^{\text{ext}} = \int_{-\infty}^{\infty} d\omega\, \omega a_\omega^\dagger a_\omega,$$

(17)

and which has the initial condition $\rho_{SE}(0) = \rho_S(0) \otimes |0\rangle_E \langle 0|$. This transformed initial condition is now far more amenable to simulation as the environment is now described by a pure, single-configuration wave function, rather than a statistical mixed state, and so no statistical sampling is required to capture the effects of temperature on the reduced dynamics!

Analysing the effective spectral density of Equation (15), it can be seen that the new extended environment has thermal detailed balance between absorption and emission processes encoded in the ratio of the coupling strengths to the positive and negative modes in the extended Hamiltonian (see Figure 1c), as opposed to the operator statistics of a thermally occupied state of the original, physical mode, i.e.

$$\frac{J_\beta(\omega)}{J_\beta(-\omega)} = \frac{\langle a_\omega a_\omega^\dagger \rangle_\beta}{\langle a_\omega^\dagger a_\omega \rangle_\beta} = e^{\beta\omega}$$

(18)

Indeed, from the system's point of view, there is no difference between the absorption of a quantum from a thermally occupied, positive energy bath mode and the creation (emission) of an excitation into an unoccupied, negative energy, bath mode. The extension to negative frequencies essentially allows the process whereby the system would be heated by the environment (absorbing pre-existing quanta in the thermal bath) to be mimicked by spontaneous emission into a negative energy vacuum of states, as shown in Figure 1b.

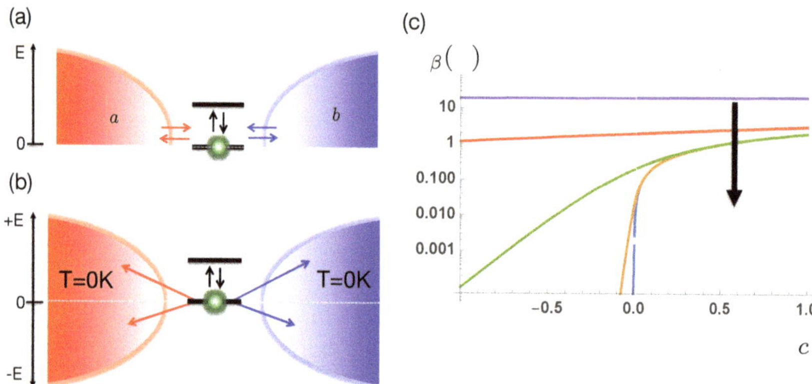

Figure 1. (**a**) Two-level system (TLS) is coupled to two environments (a, b) with inverse temperatures β_a and β_b. (**b**) The reduced state dynamics of the TLS can be obtained from a zero-temperature simulation of an extended environment containing negative frequency excitation modes and temperature-dependent couplings. (**c**) The effective spectral density $J_\beta(\omega)$ encodes the principle of detailed balance for absorption and emission of quanta between thermal transitions in the TLS. For the Ohmic spectral density considered in this article, $J_\beta(\omega)$ becomes flat over the entire range $[-\omega_c, \omega_c]$ as the temperature increases (β decreases). The plots shown are for $\omega_c\beta = 0.1$ (Purple), $\omega_c\beta = 1$ (Red), $\omega_c\beta = 10$ (Green), $\omega_c\beta = 50$ (Yellow) and $\omega_c\beta = 100$ (Blue).

In fact, the equivalence between these two environments goes beyond the reduced system dynamics as there exists a unitary transformation which links the extended environment to the original thermal environment. This means that one is able to reverse the transformation and calculate thermal expectations for the actual bosonic bath such as

$\langle a_\omega^\dagger(t) a_\omega(t) \rangle_\beta$. This is particularly useful for molecular systems in which environmental (vibrational) dynamics are also important observables that report on the mechanisms and pathways of physio-chemical transformations [38–40]. In this article, we will use this capability later to look at the non-equilibrium heat flows between the TLS and its environments. This is a major advantage of many-body wave function approaches, as full information about the environment is available, cf. effective master equation descriptions which are obtained after averaging over the environmental state.

2.3. Chain Mapping and Chain Coefficients

Following this transformation a further step is required to facilitate efficient simulation of the many-body system+environment wave-function. This is to apply a unitary transformation to the bath modes which converts the star-like geometry of H_I^{ext} into a chain-like geometry, thus allowing the use of Matrix-Product-State (MPS) methods [10,41,42] (see Figure 2). We thus define new modes $c_n^{(\dagger)} = \int_{-\infty}^\infty U_n(\omega) a_\omega^{(\dagger)}$, known as chain modes, via the unitary transformation $U_n(\omega) = \sqrt{J_\beta(\omega)} p_n(\omega)$ where $p_n(\omega)$ are orthonormal polynomials with respect to the measure $d\omega J_\beta(\omega)$. Thanks to the three term recurrence relations associated with all orthonormal polynomials $p_n(\omega)$ [41], only one of these new modes, $n = 1$, will be coupled to the system, while all other chain modes will be coupled only to their nearest neighbours [41]. Our interaction and bath Hamiltonians thus become

$$H_I^{\text{chain}} = \kappa A_S(c_1 + c_1^\dagger),$$

$$H_E^{\text{chain}} = \sum_{n=1}^\infty \epsilon_n c_n^\dagger c_n + \sum_{n=1}^\infty (t_n c_n^\dagger c_{n+1} + h.c). \tag{19}$$

The chain coefficients appearing in Equation (19) are related to the three-term recurrence parameters of the orthonormal polynomials and can be computed using standard numerical techniques [41]. Since the initial state of the bath was the vacuum state, it is unaffected by the chain transformation. We briefly note the evolution of the asymptotic values of the chain parameters, as illustrated in Figure 3. For a smooth spectral density with a hard cut-off, it is has been rigorously proven that $\epsilon_n \to \omega_c/2, t_n \to \omega_c/4$ as $n \to \infty$ [41]. Figure 3 shows the dramatic changes in these asymptotic values as the temperature is increased, which—from the numerical results—appear to be $\epsilon_n \to 0, t_n \to \omega_c/2$ as $n \to \infty$ and $\beta \to 0$. This can be naturally understood from the behaviour of the effective spectral functions $J_\beta(\omega)$ with increasing temperature, as illustrated in Figure 1c. The spectral functions become symmetric and have finite values over the whole domain $[-\omega_c, \omega_c]$. The asymptotic spectrum of the chain modes thus has a bandwidth of $2\omega_c$ centred on $\omega = 0$, which, for a uniform hopping chain, requires the numerically observed asymptotic chain parameters. In the particular case of the Ohmic environment at high temperatures, it can easily be seen that $J_\beta(\omega)$ tends to a constant and so will have a chain representation derived from the classical Legendre polynomials [41].

We have thus arrived at a formulation of the problem of finite-temperature open systems in which the many-body environmental state is initialised as a pure product of trivial ground states, whilst the effects of thermal fluctuations and populations are encoded in the Hamiltonian chain parameters and system-chain coupling. These parameters must be determined once for each temperature but—in principle—the actual simulation of the many-body dynamics is now no more complex than a zero-temperature simulation!

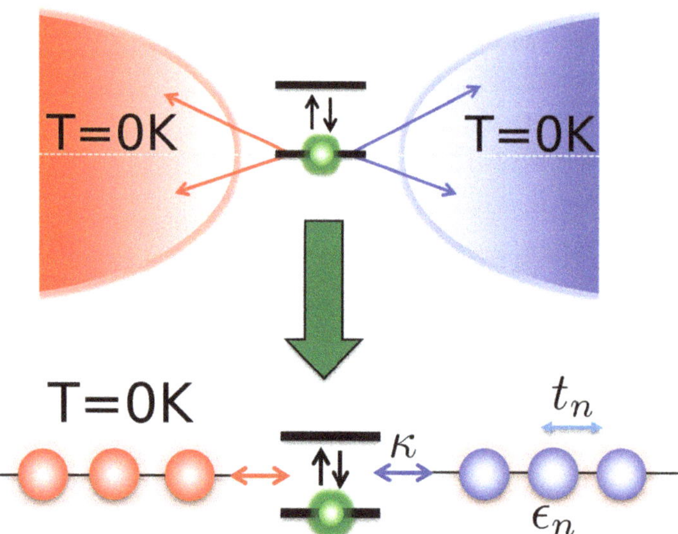

Figure 2. The positive and negative energy modes of each extended environment are mapped onto 1D chains with nearest-neighbour hopping, each coupled by their first site to the TLS with coupling strength κ. The chain parameters ϵ_n and t_n are determined such that the eigen-modes of the chains are the original modes of the extended environments. The 1D geometry of the transformed system and the fact that the chain modes all start in their vacuum states, means the system-environment state can be described by a single (pure) matrix product states (MPS).

2.4. Spin-Boson Model Across the Complete $\alpha - \beta$ Space

In this section, we numerically verify that the finite-temperature approach set out in Sections 2.2 and 2.3 captures the correct non-perturbative behaviour in the single-bath spin-boson model. This will be illustrated with a few explicit examples, but the key result of this section is the creation of a comprehensive dataset for the Ohmic spin-boson model that allows arbitrary TLS initial conditions to be propagated in real-time and over a large area of $\alpha - \beta$ space. This dataset has been made freely available online in citable form and can be used to benchmark other methods and applications [43].

Figure 4a,b shows the temporal decay of an initially polarised spin $\langle \sigma_z(0) \rangle = +1$ towards thermal equilibrium for varying coupling strengths α and inverse temperatures β. The TLS energy splitting was $\omega_0 = 0.2\omega_c$. The key result in Figure 4b is the dependence of the thermalized spin polarization at long times. In a simple, perturbative rate equation treatment, this final polarization would be set by the energy gap ω_0 and the temperature, according to the Gibbs-Boltzmann distribution

$$\langle \sigma_z \rangle_\beta = -\frac{\left(1 - e^{-\beta\omega_0}\right)}{\left(1 + e^{-\beta\omega_0}\right)}. \tag{20}$$

The coupling strength α would only alter the rate at which this thermal distribution is reached. However, Figure 4a shows a growing dependence of the final polarization on the coupling strength, suggesting a non-perturbative effect. This is indeed the case: strong coupling leads to polaron formation and non-perturbative renormalisation of the TLS energy gap ω_0. According to the variational theory of Silbey and Harris [44], the renormalized gap ω_r is approximately given by

$$\omega_r = \omega_0 \left(\frac{\omega_0}{\omega_c}\right)^{\frac{\alpha}{1-\alpha}}, \tag{21}$$

in the so-called scaling limit in which ω_c is much larger than all other energy scales in the problem. This renormalisation is highly non-perturbative, and can completely close the TLS energy gap at a critical coupling $\alpha_c = 1$ [1]. Replacing ω_0 with the the renormalized energy gaps in Equation (20), $\langle \sigma_z \rangle_\beta$ is given

$$\langle \sigma_z \rangle_\beta = -\left(\frac{\omega_0}{\omega_c}\right)^{\frac{\alpha}{1-\alpha}} \frac{(1 - e^{-\beta\omega_r})}{(1 + e^{-\beta\omega_r})}.$$

(22)

where the prefactor in Equation (22) accounts for the suppressed expectation values of σ_z in the polaronic eigenbasis.

Figure 5 shows this analytical prediction as a function of temperature, compared to the results extracted from the real-time dynamics. As mentioned above, most analytical predictions for the SBM are obtained deep in the scaling limit, while numerical results necessarily involve only moderately large values of ω_c. When comparing results, it is common in the literature to evaluate analytical expressions with a re-scaled coupling strength $\tilde{\alpha} = c\alpha$ to account for this [45–47], which we have applied in Figure 5. We found that a constant factor $c = 0.66$ gave excellent agreement across the parameter space for both one and two-bath SBMs, as shown in the inset of Figure 5.

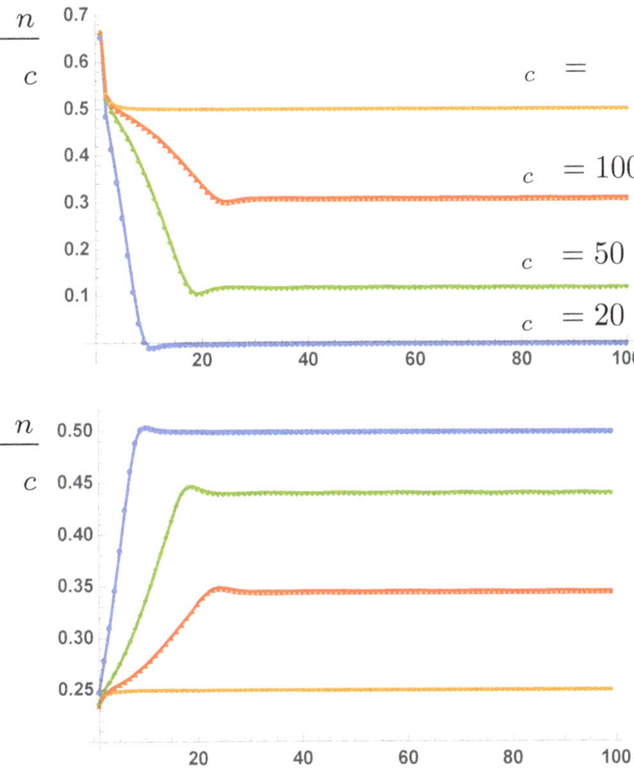

Figure 3. Site energies e_n and hopping amplitudes t_n as a function of chain distance n at different environment temperatures.

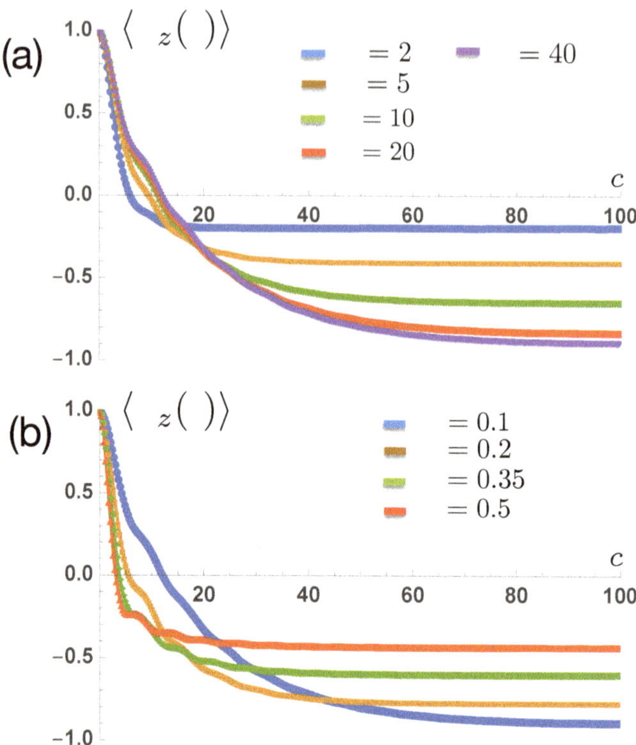

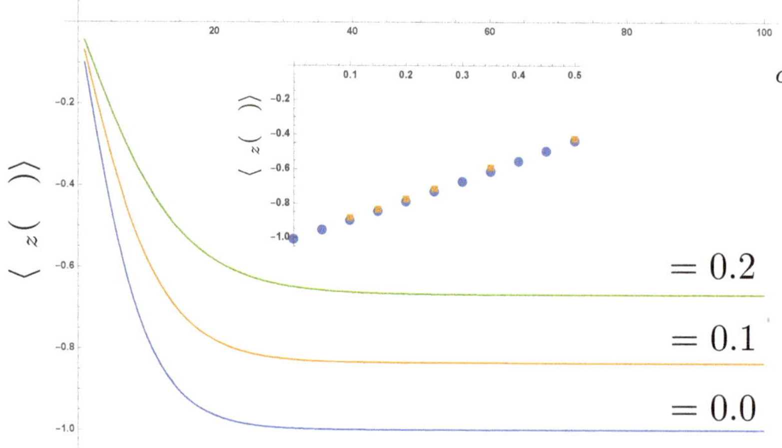

Figure 4. Relaxation of spin polarization as a function of time for (**a**) different temperatures and $\alpha = 0.1$ and (**b**) different coupling strengths with a fixed $\omega_c \beta = 100$.

Figure 5. Analytical prediction of thermal steady state spin polarization as a function of inverse temperature $\omega_c \beta$. Inset compares these predictions with steady state values extracted from the real-time dynamics shown in Figure 4. A re-scaled coupling strength $\tilde{\alpha} = c\alpha$ with $c = 0.66$ has been applied when evaluating the anaytical formula (see main text).

As a final set of observations in this section, we now look at the behaviour of the environment. In Figure 6 we present the occupations of the bath modes in the extended spectral representation used to account for finite temperatures. As anticipated in our discussion in Section 2.2, we find that at low temperatures, the energy released from the decay of the spin is absorbed by modes with positive frequencies matching the TLS energy gap ω_0. As the temperature increases, peaks appear at negative frequencies, corresponding to the excitation of these modes due to 'heating' of the TLS, i.e., the TLS is thermally excited and removes energy from the environment. As a function of temperature, the ratio of the positive and negative occupations is a very close fit to $e^{\beta\omega_0}$, as expected from detailed balance. However, due to the presence of negative frequency modes, we find that the populations in both the positive and negative frequency modes grow indefinitely during the simulation time, as shown in the inset of Figure 6. The difference of these growing populations plateaus at a finite value, corresponding to the thermal occupation of the physical positive-frequency mode, but care must be taken to get converged results for long-time (steady state) quantities due to the expanding local Hilbert spaces needed for the environment modes in the simulations.

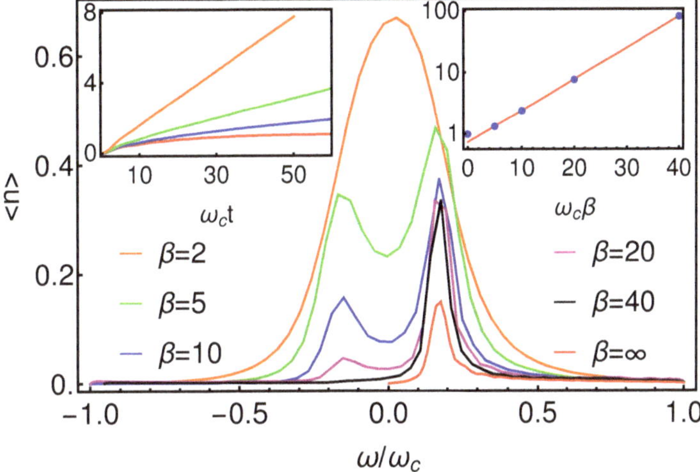

Figure 6. Long-time occupations of the modes of the extended environment, following the thermalization of the TLS at various temperatures. Inset shows the total number of quanta in the environment as a function of time (left) for different temperatures. This population grows indefinitely at finite temperatures. The inset (right) shows the long-time ratio of the peak heights in each curve of the main figure. These give a very good fit to the exponential dependence expected for absorption and emission rates obeying detailed balance.

Figure 7 shows the behaviour of the von-Neumann entropy obtained by bi-partitioning the total N-site system-environment chain into chains of size n and $N - n$ and computing the singular values of either of the subsystems' reduced density matrices [48]. The entanglement entropy directly reports on the size of the bond-dimensions required to represent the state accurately in the MPS format, and our results show that this entropy also grows continuously during the simulation. There is also a clear asymmetry in the rate of spreading and magnitudes of entanglement, with correlations between sites in the hot environment growing much faster. Again, these growing numerical resources for finite temperature simulations should be handled with care, and we shall take this up again in Section 3.

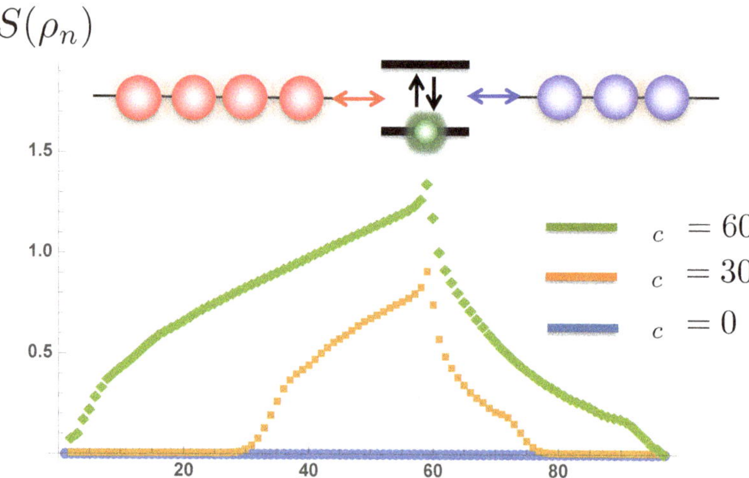

Figure 7. The time evolution of the von Neumann entanglement entropy for each bi-partition of the system-environment chain at site n. The TLS is located in this example at site $n = 59$, with the hot bath corresponding to sites $1 - 58$ and remaining sites representing the cold bath.

2.5. Non-Equilibrium Heat Flows

In this section we simulate the non-equilibrium dynamics of the TLS connected to two environments at different temperatures. For clarity we will designate environment a as the 'hot' environment and b as the 'cold' one, using suffixes 'h' and 'c', respectively. We note here that this elementary class of two-environment models has both wide-ranging practical applications—such as studying heat and charge transfer in nano-devices and molecules [6,49,50], as well as being of fundamental relevance for quantum thermodynamics, decoherence, and non-equilibrium steady states [47,51–54].

Figure 8 shows the real-time excitation of a TLS initially prepared in its ground state when connected at $t = 0$ to the cold environment with fixed $\omega_0 \beta_c = 100$ and the 'hot' environment at different temperatures. Figure 9 shows the steady-state spin polarization as a function of the temperature difference between the hot and cold baths. To understand the basic features of the steady state, let's consider a perturbative set of rate equations for the population of the spin-up level $P_\uparrow(t)$. Assuming that the rates of absorption and emission from each bath of TLS obey detailed balance, the dynamics of $P_\uparrow(t)$ can be obtained from the equation

$$\frac{dP_\uparrow(t)}{dt} = -\Gamma P_\uparrow(t)(n_c + n_h + 2) + \Gamma(1 - P_\uparrow(t))(n_c + n_h), \tag{23}$$

where $n_i = [\exp(\omega_0 \beta_i) - 1]^{-1}$ By finding the steady state population $P_\uparrow(\infty)$, the non-equilibrium value of the spin polarisation $\langle \sigma_z^{ab}(\infty) \rangle$ then takes the simple form

$$\langle \sigma_z^{ab}(\infty) \rangle = \frac{-1}{2(n_c + n_h + 1)}. \tag{24}$$

Once again, if renormalization effects are included, the agreement between the analytical predictions is very good, as can be seen in Figure 9. Indeed, for the lowest temperatures, the spin dynamics are entirely due to renormalization effects, as thermal occupation of the excited level is negligible.

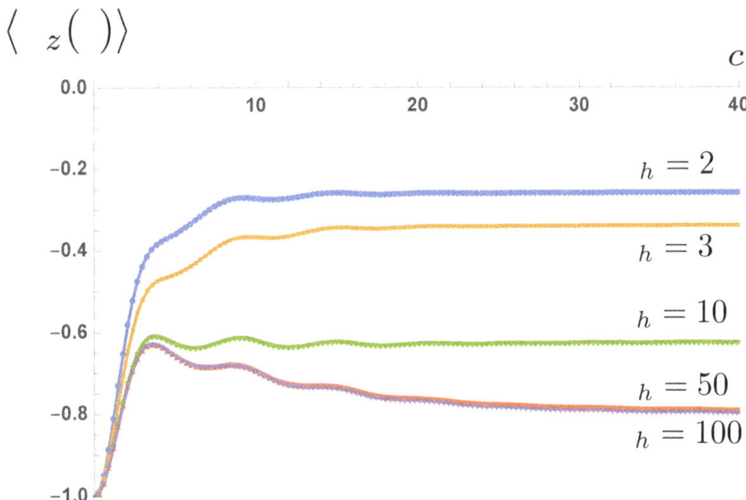

Figure 8. Non-equilibrium relaxation of spin polarization as a function of time for fixed cold bath temperature and varying hot bath temperatures.

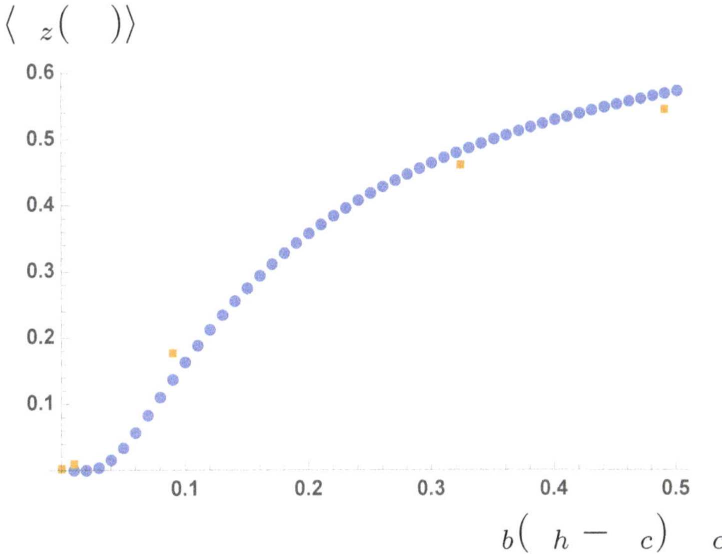

Figure 9. The change in $\langle \sigma_z(\infty) \rangle$ as a function of the temperature difference between the baths (T_c is kept constant). Analytical predictions are shown as dots, numerical data points as squares.

Interestingly, these two-bath results also reveal an intriguing non-additive effect due to the coupling to two environments. The subject of non-additivity of environmental interactions has recently attracted attention due to the role of multiple environments in a wide range of 'active' quantum machines, such as the conversion of ambient solar energy in room-temperature (phonon-coupled) devices [15,16,47], and also the highly co-operative actions of different types of vibrational motion in molecular photo physics [14]. In the present case, the non-additive effects appear in the polaronic renormalisation, which is mostly clearly seen in the case when the two baths have the same temperature. This situation is indistinguishable from a coupling to a single bath with twice the coupling strength. The renormalisation can thus be obtained from Equation (21) with the replacement $\alpha \to 2\alpha$. However, the renormalisation arises from the overlap of the displaced mode wave functions that are 'fast' enough to co-tunnel with the TLS as it transitions between $\langle \sigma_x \rangle = \pm 1$ [44,45], and in an additive approximation the renormalization would be simply be the product of the individual overlaps for each environment. However, as is clear form Equation (21), this doubling of the coupling does not lead to a simple exponential doubling of the renormalization, but instead leads to a nonlinear suppression of the energy gap according to the exponent $2\alpha/(1 - 2\alpha)$.

Exploiting the access to the environmental state, we now show the transient dynamics of the heat flow in the two baths during the establishment of the TLS steady state. We define the following operators

$$\hat{J}_c = \hat{\sigma}_y \otimes (\hat{A}_0^\dagger + \hat{A}_0), \tag{25}$$

and

$$\hat{J}_h = \hat{\sigma}_y \otimes (\hat{B}_0^\dagger + \hat{B}_0), \tag{26}$$

which measure the heat flux from the spin to baths a and b respectively. The operators $\hat{A}_0^{(\dagger)}(\hat{B}_0^{(\dagger)})$ refer to the creation and anihilation operators of the first site of chain a(b), i.e. the site coupled to the TLS. Representative heat flows are shown in Figure 10 for large and zero differences in the bath temperatures. In both cases, the initial dynamics involve heating from both hot and cold environments, as the spin is initially in a pure $(T = 0K)$ ground state. As the dynamical steady state of the spin is obtained, a net heat current appears from the hot to cold environment. This heat current vanishes as the temperature difference of the baths is reduced, as we would expect. From the long-time solution of the Pauli master equation given in Equation (23), the steady-state heat flux from the hot to cold environment can be shown to be

$$J = \Gamma \frac{n_h - n_c}{1 + n_h + n_c}, \tag{27}$$

and this is plotted alongside our numerical data in Figure 11. The simulations correctly capture the essentially non-linear behaviour of heat flow through the quantum 'heat leak' TLS, although a linear regime where Fourier's law of heat flow appears to hold can be clearly observed before the flows saturate for large temperature differences.

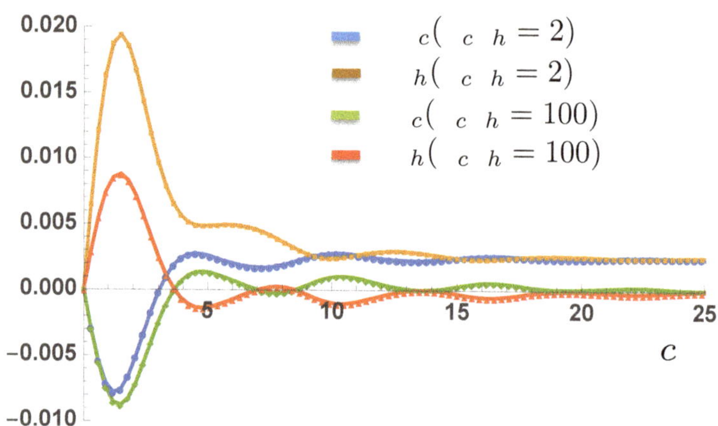

Figure 10. Heat flows into the cold bath (J_c) and out of the hot bath (J_h) as a function of time for varying hot bath temperatures and a fixed $\omega_c\beta_c = 100$.

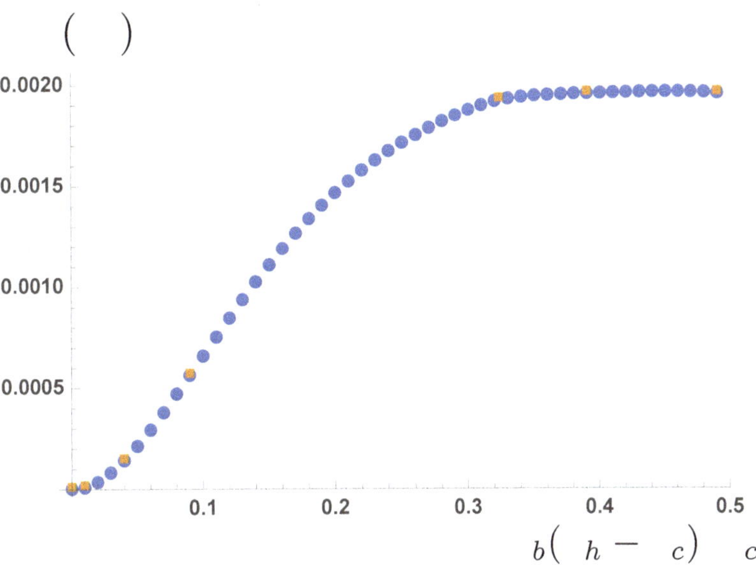

$$\beta_c\left(\beta_h - \beta_c\right)/\beta_c$$

Figure 11. Net steady-state heat flux through the two-level system as a function of temperature difference for a fixed $\beta_c = 100$. Data extracted from MPS simulations (yellow squares) is compared with the analytical expression in the main text (blue dots).

3. Discussion

The results presented in Section 2 demonstrate that accurate reduced system behaviour in the spin-boson model can be obtained in the presence of both a single or two finite-temperature environments. Non-perturbative effects related to system-bath entanglement (polaronic dressing) are captured in transient relaxation and non-equilibrium steady states, and we have shown how the T-TEDOPA transformation provides direct information related to the energy and entanglement entropy flows in the environment. All of these results were obtained from pure wave function evolution of an initial zero-temperature (vacuum) state, and the onerous numerical cost of having to sample over a thermal distribution of initial states was entirely avoided.

However, we did note that the numerical resources required to obtain these results grew in an unbounded way as a function of simulation time. In the case of the one-bath SBM, Figure 6 shows that the total number of bosonic excitation grows approximately linearly in time and the growth rate increases with the bath temperature. The main panel showing the populations of the environment in frequency space shows that this growth is the result of growing populations at frequencies $\approx \pm\omega_0$. In Section 2.2 we made the observation that creating an excitation in a negative frequency mode allows the TLS to be excited with overall conservation of energy, and this is the process that accounts for the 'heating' expected of a finite-temperature bath. The constant growth of excitations in the environment can be seen to arise from the constant cycling of the heating process sketched in Figure 12 (a similar cycle for cooling also generates a net population of excitations). Here the creation of a negative frequency excitation (or hole) excites the TLS and then is de-excited by the creation of an excitation in the positive frequency environment.

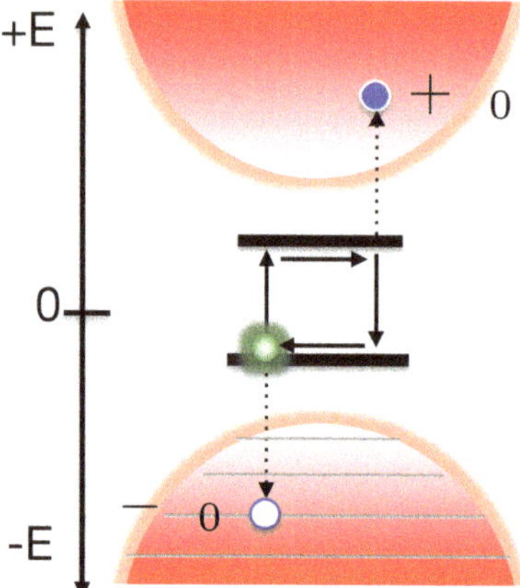

Figure 12. Due to the unbounded bosonic nature of the negative-frequency environment, thermal transitions within the TLS lead to a constant creation of correlated, particle-hole-like excitations in both environments with the same absolute energy ω_0.

Interestingly, this pair creation goes beyond populations: the dynamics of thermalization entangles the positive and negative frequency environments. This is perhaps unsurprising in the context of the thermofield theory of De Vega et al. where the thermal entanglement properties of two-mode squeezed states are used to create an effective finite temperature environment from two zero-temperature baths [55]. However, for our open-system problem, it should be kept in mind that the 'dynamics' of the positive and—especially—the negative modes really only provide insight into the internal workings of the simulation. The modes and their populations are proxy (non-physical) degrees of freedom used to provide vacuum fluctuations that mimic the physical interactions of the system with a strictly positive-frequency harmonic bath at finite-temperature. However, a hopefully fruitful and more physical connection between the behaviour of the artifically extended environment inT-TEDOPA can be made to very recent developments in the theory of MPS and tensor networks for fermionic quantum transport. Here, non-equilibrium particle flows between reservoirs at different chemical potentials lead to the constant creation of entangled particle-hole pairs, leading to the exponential-in-time growth of MPS bond dimensions. However, Rams and Zwolak have recently demonstrated that a change in basis used for certain fermionic transport simulations can greatly suppress the rapid growth of numerical resources [56], and it would be very interesting to see how this might translate—or might to some extent already be implemented—in our current approach to bosonic heat flow problems. Finally, we also point out that rapid growth of bond dimensions and entanglement in non-equilibrium systems is potentially a problem for 1TDVP simulations, as these proceed at fixed bond-dimensions. Choosing large bond-dimensions may allow one to reach long times, but much of the simulation is likely to run slowly, as it will be using far more resources than are necessary for most of the time. In a recent development, Dunnett and Chin have proposed an adaptive version of 1TDVP that is able to change bond-dimensions during the course of a single simulation run, allowing the necessary resources to be deployed as needed [57]. We thus conclude that recent insights and computational development have opened a whole new domain of finite and multiple-temperature open system problems for wave function techniques, and that creating numerically efficient finite-temperature simulations will inspire further progress in tensor network theory, as applied to open systems.

4. Materials and Methods

All numerical results were obtained using software packages that are available at https://github.com/angusdunnett/MPSDynamics. The benchmark data for the Ohmic Spin-Boson Model can be found at [43].

Author Contributions: Conceptualisation, A.W.C.; methodology, A.J.D. & A.W.C.; software, A.J.D.; writing—original draft preparation, A.W.C. & A.J.D. All authors have read and agreed to the published version of the manuscript.

Funding: ADJ Acknowledges support from Ecole Doctorale Physique en Ile-de-France (EDPIF ED564. A.W.C. is partly supported by ANR project No. 195608/ACCEPT.

Institutional Review Board Statement: Not applicable.

Informed Consent Statement: Not applicable.

Data Availability Statement: Data sharing not applicable.

Conflicts of Interest: The authors declare no conflict of interest.

References

1. Weiss, U. *Quantum Dissipative Systems*; World Scientific: Singapore, 2012; Volume 13. [CrossRef]
2. Breuer, H.P.; Petruccione, F. *The Theory of Open Quantum Systems*; Oxford University Press on Demand: New York, NY, USA, 2002.
3. Acín, A.; Bloch, I.; Buhrman, H.; Calarco, T.; Eichler, C.; Eisert, J.; Esteve, D.; Gisin, N.; Glaser, S.J.; Jelezko, F.; et al. The quantum technologies roadmap: a European community view. *New J. Phys.* **2018**, *20*, 080201. [CrossRef]
4. Gemmer, J.; Michel, M.; Mahler, G. *Quantum Thermodynamics: Emergence of Thermodynamic Behavior within Composite Quantum Systems*; Springer: Berlin/Heidelberg, Germany, 2009; Volume 784.
5. Kosloff, R.; Levy, A. Quantum Heat Engines and Refrigerators: Continuous Devices. *Annu. Rev. Phys. Chem.* **2014**, *65*, 365–393. [CrossRef] [PubMed]
6. Benenti, G.; Casati, G.; Saito, K.; Whitney, R. Fundamental aspects of steady-state conversion of heat to work at the nanoscale. *Phys. Rep.* **2017**, *694*, 1–124. [CrossRef]
7. Elenewski, J.E.; Gruss, D.; Zwolak, M. Communication: Master equations for electron transport: The limits of the Markovian limit. *J. Chem. Phys.* **2017**, *147*, 151101. [CrossRef] [PubMed]
8. Thoss, M.; Evers, F. Perspective: Theory of quantum transport in molecular junctions. *J. Chem. Phys.* **2018**, *148*, 030901. [CrossRef]
9. Ishizaki, A.; Fleming, G.R. Unified treatment of quantum coherent and incoherent hopping dynamics in electronic energy transfer: Reduced hierarchy equation approach. *J. Chem. Phys.* **2009**, *130*, 234111. [CrossRef]
10. Chin, A.; Prior, J.; Rosenbach, R.; Caycedo-Soler, F.; Huelga, S.F.; Plenio, M.B. The role of non-equilibrium vibrational structures in electronic coherence and recoherence in pigment–protein complexes. *Nat. Phys.* **2013**, *9*, 113–118. [CrossRef]
11. Smith, S.L.; Chin, A.W. Ultrafast charge separation and nongeminate electron–hole recombination in organic photovoltaics. *Phys. Chem. Chem. Phys.* **2014**, *16*, 20305–20309. [CrossRef]
12. Oviedo-Casado, S.; Prior, J.; Chin, A.; Rosenbach, R.; Huelga, S.; Plenio, M. Phase-dependent exciton transport and energy harvesting from thermal environments. *Phys. Rev. A* **2016**, *93*, 020102. [CrossRef]
13. Chin, A.; Mangaud, E.; Atabek, O.; Desouter-Lecomte, M. Coherent quantum dynamics launched by incoherent relaxation in a quantum circuit simulator of a light-harvesting complex. *Phys. Rev. A* **2018**, *97*, 063823. [CrossRef]
14. Schröder, F.A.; Turban, D.H.; Musser, A.J.; Hine, N.D.; Chin, A.W. Tensor network simulation of multi-environmental open quantum dynamics via machine learning and entanglement renormalisation. *Nat. Commun.* **2019**, *10*, 1062.
15. Maguire, H.; Iles-Smith, J.; Nazir, A. Environmental nonadditivity and franck-condon physics in nonequilibrium quantum systems. *Phys. Rev. Lett.* **2019**, *123*, 093601. [CrossRef] [PubMed]
16. Wertnik, M.; Chin, A.; Nori, F.; Lambert, N. Optimizing co-operative multi-environment dynamics in a dark-state-enhanced photosynthetic heat engine. *J. Chem. Phys.* **2018**, *149*, 084112. [CrossRef] [PubMed]
17. Del Pino, J.; Schröder, F.A.; Chin, A.W.; Feist, J.; Garcia-Vidal, F.J. Tensor network simulation of polaron-polaritons in organic microcavities. *Phys. Rev. B* **2018**, *98*, 165416. [CrossRef]
18. Strathearn, A.; Kirton, P.; Kilda, D.; Keeling, J.; Lovett, B.W. Efficient non-Markovian quantum dynamics using time-evolving matrix product operators. *Nat. Commun.* **2018**, *9*, 3322. [CrossRef]
19. Topaler, M.; Makri, N. Quantum rates for a double well coupled to a dissipative bath: Accurate path integral results and comparison with approximate theories. *J. Chem. Phys.* **1994**, *101*, 7500–7519. [CrossRef]
20. Prior, J.; Chin, A.W.; Huelga, S.F.; Plenio, M.B. Efficient simulation of strong system-environment interactions. *Phys. Rev. Lett.* **2010**, *105*, 050404. [CrossRef]
21. Somoza, A.D.; Marty, O.; Lim, J.; Huelga, S.F.; Plenio, M.B. Dissipation-Assisted Matrix Product Factorization. *Phys. Rev. Lett.* **2019**, *123*, 100502. [CrossRef]

22. Lindner, C.J.; Kugler, F.B.; Meden, V.; Schoeller, H. Renormalization group transport theory for open quantum systems: Charge fluctuations in multilevel quantum dots in and out of equilibrium. *Phys. Rev. B* **2019**, *99*, 205142. [CrossRef]
23. Wang, H.; Shao, J. Quantum Phase Transition in the Spin-Boson Model: A Multilayer Multiconfiguration Time-Dependent Hartree Study. *J. Phys. Chem. A* **2019**, *123*, 1882–1893. [CrossRef]
24. Haegeman, J.; Lubich, C.; Oseledets, I.; Vandereycken, B.; Verstraete, F. Unifying time evolution and optimization with matrix product states. *Phys. Rev. B* **2016**, *94*, 165116. [CrossRef]
25. Haegeman, J.; Cirac, J.I.; Osborne, T.J.; Pižorn, I.; Verschelde, H.; Verstraete, F. Time-Dependent Variational Principle for Quantum Lattices. *Phys. Rev. Lett.* **2011**, *107*, 070601. [CrossRef] [PubMed]
26. Schröder, F.A.Y.N.; Chin, A.W. Simulating open quantum dynamics with time-dependent variational matrix product states: Towards microscopic correlation of environment dynamics and reduced system evolution. *Phys. Rev. B* **2016**, *93*, 075105. [CrossRef]
27. Gonzalez-Ballestero, C.; Schröder, F.A.Y.N.; Chin, A.W. Uncovering nonperturbative dynamics of the biased sub-Ohmic spin-boson model with variational matrix product states. *Phys. Rev. B* **2017**, *96*, 115427. [CrossRef]
28. Tamascelli, D.; Smirne, A.; Lim, J.; Huelga, S.F.; Plenio, M.B. Efficient simulation of finite-temperature open quantum systems. *Phys. Rev. Lett.* **2019**, *123*, 090402.
29. Tamascelli, D.; Smirne, A.; Huelga, S.; Plenio, M. Nonperturbative Treatment of non-Markovian Dynamics of Open Quantum Systems. *Phys. Rev. Lett.* **2018**, *120*, 030402. [CrossRef]
30. Wilhelm, F.; Kleff, S.; Von Delft, J. The spin-boson model with a structured environment: a comparison of approaches. *Chem. Phys.* **2004**, *296*, 345–353. [CrossRef]
31. Schulze, J.; Kuhn, O. Explicit correlated exciton-vibrational dynamics of the FMO complex. *J. Phys. Chem. B* **2015**, *119*, 6211–6216. [CrossRef]
32. Mendive-Tapia, D.; Mangaud, E.; Firmino, T.; de la Lande, A.; Desouter-Lecomte, M.; Meyer, H.D.; Gatti, F. Multidimensional quantum mechanical modeling of electron transfer and electronic coherence in plant cryptochromes: The role of initial bath conditions. *J. Phys. Chem. B* **2018**, *122*, 126–136. [CrossRef]
33. May, V.; Kühn, O. *Charge and Energy Transfer Dynamics in Molecular Systems*; John Wiley & Sons: Hoboken, NJ, USA, 2008.
34. Mukamel, S. *Principles of Nonlinear Optical Spectroscopy*; Oxford University Press: New York, NY, USA, 1995; Volume 6.
35. Alvermann, A.; Fehske, H. Sparse polynomial space approach to dissipative quantum systems: Application to the sub-ohmic spin-boson model. *Phys. Rev. Lett.* **2009**, *102*, 150601. [CrossRef]
36. Binder, R.; Burghardt, I. First-principles quantum simulations of exciton diffusion on a minimal oligothiophene chain at finite temperature. *Faraday Discuss.* **2019**, *221*, 406–427. [CrossRef] [PubMed]
37. Jiang, T.; Li, W.; Ren, J.; Shuai, Z. Finite Temperature Dynamical Density Matrix Renormalization Group for Spectroscopy in Frequency Domain. *T J. Phys. Chem. Lett.* **2020**, *11*, 3761–3768.
38. Musser, A.J.; Liebel, M.; Schnedermann, C.; Wende, T.; Kehoe, T.B.; Rao, A.; Kukura, P. Evidence for conical intersection dynamics mediating ultrafast singlet exciton fission. *Nat. Phys.* **2015**, *11*, 352–357. [CrossRef]
39. Schnedermann, C.; Lim, J.M.; Wende, T.; Duarte, A.S.; Ni, L.; Gu, Q.; Sadhanala, A.; Rao, A.; Kukura, P. Sub-10 fs time-resolved vibronic optical microscopy. *J. Phys. Chem. Lett.* **2016**, *7*, 4854–4859. [CrossRef]
40. Schnedermann, C.; Alvertis, A.M.; Wende, T.; Lukman, S.; Feng, J.; Schröder, F.A.; Turban, D.H.; Wu, J.; Hine, N.D.; Greenham, N.C.; et al. A molecular movie of ultrafast singlet fission. *Nat. Commun.* **2019**, *10*, 4207.
41. Chin, A.W.; Rivas, Á.; Huelga, S.F.; Plenio, M.B. Exact mapping between system-reservoir quantum models and semi-infinite discrete chains using orthogonal polynomials. *J. Math. Phys.* **2010**, *51*, 092109. [CrossRef]
42. Prior, J.; de Vega, I.; Chin, A.W.; Huelga, S.F.; Plenio, M.B. Quantum dynamics in photonic crystals. *Phys. Rev. A* **2013**, *87*, 013428. [CrossRef]
43. Chin, A.; Dunnett, A. Real-time benchmark dynamics of the Ohmic Spin- Boson Model computed with Time-Dependent Variational Matrix Product States. (TDVMPS) coupling strength and temperature parameter space. *Zenodo* **2020**. [CrossRef]
44. Silbey, R.; Harris, R.A. Variational calculation of the dynamics of a two level system interacting with a bath. *J. Chem. Phys.* **1984**, *80*, 2615–2617. [CrossRef]
45. Blunden-Codd, Z.; Bera, S.; Bruognolo, B.; Linden, N.O.; Chin, A.W.; Von Delft, J.; Nazir, A.; Florens, S. Anatomy of quantum critical wave functions in dissipative impurity problems. *Phys. Rev. B* **2017**, *95*, 085104. [CrossRef]
46. Florens, S.; Freyn, A.; Venturelli, D.; Narayanan, R. Dissipative spin dynamics near a quantum critical point: Numerical renormalization group and Majorana diagrammatics. *Phys. Rev. B* **2011**, *84*, 155110. [CrossRef]
47. Bruognolo, B.; Weichselbaum, A.; Guo, C.; von Delft, J.; Schneider, I.; Vojta, M. Two-bath spin-boson model: Phase diagram and critical properties. *Phys. Rev. B* **2014**, *90*, 245130. [CrossRef]
48. Nielsen, M.A.; Chuang, I. Quantum computation and quantum information. *Am. J. Phys.* **2002**, *70*, 558. [CrossRef]
49. Dubi, Y.; Di Ventra, M. *Colloquium*: Heat flow and thermoelectricity in atomic and molecular junctions. *Rev. Mod. Phys.* **2011**, *83*, 131–155. [CrossRef]
50. Dhar, A. Heat transport in low-dimensional systems. *Adv. Phys.* **2008**, *57*, 457–537. [CrossRef]
51. Guo, C.; Weichselbaum, A.; von Delft, J.; Vojta, M. Critical and Strong-Coupling Phases in One- and Two-Bath Spin-Boson Models. *Phys. Rev. Lett.* **2012**, *108*, 160401. [CrossRef] [PubMed]

52. Zhou, N.; Chen, L.; Xu, D.; Chernyak, V.; Zhao, Y. Symmetry and the critical phase of the two-bath spin-boson model: Ground-state properties. *Phys. Rev. B* **2015**, *91*, 195129. [CrossRef]
53. Segal, D.; Nitzan, A. Spin-Boson Thermal Rectifier. *Phys. Rev. Lett.* **2005**, *94*, 034301. [CrossRef]
54. Chen, T.; Balachandran, V.; Guo, C.; Poletti, D. Steady state quantum transport through an anharmonic oscillator strongly coupled to two heat reservoirs. *arXiv* **2020**, arXiv:2004.05017.
55. de Vega, I.; Bañuls, M.C. Thermofield-based chain-mapping approach for open quantum systems. *Phys. Rev. A* **2015**, *92*, 052116. [CrossRef]
56. Rams, M.M.; Zwolak, M. Breaking the Entanglement Barrier: Tensor Network Simulation of Quantum Transport. *Phys. Rev. Lett.* **2020**, *124*, 137701. [CrossRef] [PubMed]
57. Dunnett, A.J.; Chin, A.W. Dynamically Evolving Bond-Dimensions within the one-site Time-Dependent-Variational-Principle method for Matrix Product States: Towards efficient simulation of non-equilibrium open quantum dynamics. *arXiv* **2020**, arXiv:2007.13528.

Article

Excitation Dynamics in Chain-Mapped Environments

Dario Tamascelli [1,2]

[1] Dipartimento di Fisica "Aldo Pontremoli", Università degli Studi di Milano, via Celoria 16,
20133 Milano, Italy; dario.tamascelli@unimi.it
[2] Institut für Theoretische Physik and Center for Integrated Quantum Science and Technology (IQST),
Albert-Einstein-Allee 11, Universität Ulm, 89069 Ulm, Germany

Received: 26 October 2020; Accepted: 17 November 2020; Published: 19 November 2020

Abstract: The chain mapping of structured environments is a most powerful tool for the simulation of open quantum system dynamics. Once the environmental bosonic or fermionic degrees of freedom are unitarily rearranged into a one dimensional structure, the full power of Density Matrix Renormalization Group (DMRG) can be exploited. Beside resulting in efficient and numerically exact simulations of open quantum systems dynamics, chain mapping provides an unique perspective on the environment: the interaction between the system and the environment creates perturbations that travel along the one dimensional environment at a finite speed, thus providing a natural notion of light-, or causal-, cone. In this work we investigate the transport of excitations in a chain-mapped bosonic environment. In particular, we explore the relation between the environmental spectral density shape, parameters and temperature, and the dynamics of excitations along the corresponding linear chains of quantum harmonic oscillators. Our analysis unveils fundamental features of the environment evolution, such as localization, percolation and the onset of stationary currents.

Keywords: transport; open quantum systems; chain-mapping

1. Introduction

The thorough understanding of transport of energy, heat, particle, or mass in complex quantum systems is of utmost importance both from a fundamental and technological point of view. Such a relevance is witnessed by the enormous efforts invested by the scientific community over the last decades on the theoretical and experimental investigation of the unique features of transport at the quantum regime.

A variety of different topics can be put under the umbrella of quantum transport, such as efficient energy transfer and conversion in biological systems [1–5], transport in low dimensional quantum systems [6–11], quantum thermodynamics [12,13], and quantum information processing and transmission [14–17].

Open quantum systems (OQS) formalism [18,19] has been widely employed for the description of quantum transport in the, often unavoidable, presence of additional and uncontrollable degrees of freedom interacting with the system under study. The tools provided by open quantum system theory led to the derivation of fundamental results allowing to understand and control, or at least mitigate, environmental effects. Such control, for example, is of utmost importance to preserve the quantum resources, as entanglement and coherence, that could enable the development of quantum devices possibly outperforming their classical counterparts. On the other side, the analysis of certain open quantum systems has unveiled the delicate interplay between coherence and sources of decoherence, as in the paradigmatic case of energy transport in disordered lattices [2,3,16,20].

The simulation of open quantum systems, on the other hand, represents a formidable task. Even when a microscopic description of the environment surrounding a quantum system is available, the derivation of the open quantum system dynamics requires the solution of a number of differential equations that scales exponentially with the number of environmental degrees of freedom. Analytic solutions are not available but for very few cases [21–27] and numerical integration is not feasible, unless more or less severe approximations are used. Such approximations, however, may fail to capture the effects of the interaction of open systems with environments that are either structured, or evolve on time-scales comparable to those characteristic of the open system. Electronic excitation or electron transport in a vibrational environment, ubiquitous in solid state environments and bio-molecular systems [28–31], is just an example of this class of problems, which are of fundamental importance in a broad range of fields including the emergent quantum technology.

Over the last two decades, a variety of numerically exact approaches for the simulation of open quantum systems have been proposed [31]. These methods allow for the description of features that were not accurately described by approximate methods, such as the Markov, Bloch-Redfield or perturbative expansion techniques [18]. Among them we mention the hierarchical equations of motion (HEOM) [32–34], path integral methods [35–37], Dissipation-Assisted Matrix Product Factorization [38], and pseudo-modes related transformations [23,39,40].

Time Evolving Density operator with Orthogonal Polynomials (TEDOPA) [41,42] algorithm is a method for the non perturbative simulation of OQS. TEDOPA has been employed to study a variety of open quantum systems [4,41,43]. TEDOPA belongs to the class of chain-mapping techniques [41,42,44–47], based on a unitary mapping of the environmental modes onto a chain of harmonic oscillators with nearest-neighbor interactions. The main advantage of this mapping is the more local entanglement structure which allows for a straightforward application of density matrix renormalization group (DMRG) methods [48]. Moreover, the availability of bounds on the numerical errors introduced by the DMRG parametrization allows to certify the accuracy of the results generated by TEDOPA [49].

As we will show, starting from the next section, after the transformation of the environmental degrees of freedom into a linear chain of bosonic modes, the open system interacts only with the first site of the chain, where it dynamically creates (and destroys) excitations that subsequently propagate along the linear chain. A deeper understanding of excitation transport on bosonic chains obtained via the unitary chain mapping transformation of a bosonic environment can shed light on the mechanism that allows a linear structure to induce on the system the same dynamics of the original environmental configuration, where each oscillator was directly interacting with the system. The same linear structure, moreover, offers a unique point of view on the perturbations induced on the environment by the interaction with the system, since it naturally introduces a hierarchy of modes over which such perturbation propagate, or light-cone.

The paper is organized as follows. In Section 2 we briefly introduce the TEDOPA chain mapping and fix our notation. In Section 3 we discuss the dynamical features of transport on TEDOPA chain associated, respectively, to Lorentzian and Ohmic spectral densities in the single excitation subspace. In Section 4 we extend the analysis by including the interaction with the open system. Section 5 is devoted to conclusion and outlook.

2. Tedopa

Here and in what follows we consider a system interacting with a bosonic environment. The complete Hamiltonian reads ($\hbar = 1$):

$$H = H_S + H_E + H_I \tag{1}$$

$$H_E = \int d\omega\, \omega a_\omega^\dagger a_\omega$$

$$H_I = A_S \int d\omega\, h(\omega) O_\omega,$$

where H_S is the (arbitrary) free system Hamiltonian, H_E describes the free evolution of the bosonic environmental degrees of freedom, and H_I is the bilinear system -environment interaction

Hamiltonian [50], and A_S, O_ω are self-adjoint operators. This last assumption is necessary for the Thermalized-TEDOPA (T-TEDOPA) mapping [51], that we will introduce later in this section. We assume that $h(\omega)$ has finite support $[\omega_{\min}, \omega_{\max}]$, with $\omega_{\min} < \omega_{\max}$, and define the spectral density (SD), namely the positive valued function $J : [\omega_{\min}, \omega_{\max}] \to \mathbb{R}^+$, as

$$J(\omega) = h^2(\omega). \tag{2}$$

As shown in References [41,42,46] the Hamiltonian (1) can be unitarily mapped into an equivalent one through by defining a countably infinite set of new operators

$$b_n^\dagger = \int_{\omega_{\min}}^{\omega_{\max}} d\omega\, U_n(\omega) a_\omega^\dagger \tag{3}$$

$$b_n = \int_{\omega_{\min}}^{\omega_{\max}} d\omega\, U_n(\omega) a_\omega, \tag{4}$$

where

$$U_n(\omega) = h(\omega) p_n(\omega). \tag{5}$$

The operators b_n and $b_n^\dagger$ satisfy the bosonic commutation relations $[b_n, b_m^\dagger] = \delta_{nm}$; moreover, the polynomials $p_n(\omega)$ are orthogonal with respect to the measure $d\mu = J(\omega)d\omega$ and satisfy three-term recursion relations [42,46]. Thanks to these properties, the Hamiltonian (1) is mapped into the one dimensional Hamiltonian

$$H^C = H_S + \kappa_0 A_s(b_1 + b_1^\dagger) + \tag{6}$$

$$\sum_{n=1}^{+\infty} \omega_n b_n^\dagger b_n + \kappa_n(b_{n+1}^\dagger b_n + b_n^\dagger b_{n+1})$$

$$= H_S + H_I^C + H_E^C, \tag{7}$$

where, for the sake of definiteness, we have specialized the operator O_ω in (1) to $X_\omega = a_\omega + a_\omega^\dagger$. After the mapping, the system interacts with the first site of a linear (infinite) chain of bosonic modes; the system-chain interaction strength is given by [42,46]

$$\kappa_0^2 = \int_{\omega_{\min}}^{\omega_{\max}} d\omega\, J(\omega), \tag{8}$$

whereas the frequency of the first TEDOPA chain oscillator is

$$\omega_1 = \int_{\omega_{\min}}^{\omega_{\max}} d\omega\, \omega \frac{J(\omega)}{\kappa_0^2}, \tag{9}$$

namely the first moment of the normalized measure $J(\omega)/\kappa_0^2 d\omega$ on $[\omega_{\min}, \omega_{\max}]$. The remaining coefficients ω_n and κ_n are defined by the above mentioned three-terms recursion relations; while in certain cases it is possible to analytically determine their value [42], a numerically stable procedure is in general used [52,53].

For the following analysis, it is important to stress that the chain Hamiltonian H_E^C is made up of exchange terms $b_{n-1}b_n^\dagger + H.c.$ and therefore conserves the "number" operator, that is, $[N, H_E^C] = 0$ where

$$N = \bigotimes_{n=1}^{\infty} b_n^\dagger b_n. \tag{10}$$

Excitations can therefore be added or subtracted from the chain because of the interaction with the system.

The initial joint system-environment state is assumed factorized $\rho_{SE}(0) = \rho_S(0) \otimes \rho_{E,\beta}(0)$, with $\rho_{E,\beta}(0)$ a thermal state at inverse temperature $\beta = 1/k_B T$, namely

$$\rho_{E,\beta}(0) = \bigotimes_\omega \exp(-\beta \omega a_\omega^\dagger a_\omega)/\mathcal{Z}_\omega, \tag{11}$$

with $\mathcal{Z}_\omega = \mathrm{Tr}[\exp(-\beta \omega a_\omega^\dagger a_\omega)]$ the partition function. The initial state after the chain mapping is a factorized state $\rho_{SE}^C(0) = \rho_S(0) \otimes \rho_{E,\beta}^C(0)$ as well with

$$\rho_{E,\beta}^C = \exp(-\beta H_E^C)/\mathcal{Z}_E^C. \tag{12}$$

If the environment is initially at zero temperature, its initial state is the vacuum state, and the initial state of the chain is also a factorized vacuum state $|0\rangle_E^C$ (i.e., $b_k |0\rangle_E^C = 0, k = 1, 2, \ldots$): the chain contains therefore no excitations. This case provides us with the simplest setting where to analyze the transport properties of the chain corresponding to some representative spectral densities. As recently shown in Reference [51], however, by the spectral density transformation

$$J_\beta(\omega) = \frac{J'(\omega)}{2} \left[1 + \coth\left(\frac{\beta \omega}{2} \right) \right], \tag{13}$$

with $J'(\omega) = \mathrm{sign}(\omega)J(|\omega|)$, it is always possible to replace the thermal state of the original environment with the vacuum state of an extended environment, comprising negative frequencies. As the spectral density (13) is now temperature dependent, the TEDOPA chain coefficients $\omega_{n,\beta}$, $\kappa_{n,\beta}$ will be temperature dependent as well. In the following we will drop the β dependence wherever clear from the context. From now on will therefore always consider the factorized vacuum state as the initial chain state without loss of generality.

In our analysis we will consider, in particular, the Lorentzian spectral density

$$J_L(\omega) = \frac{\lambda^2}{\pi} \frac{4\gamma\Omega\omega}{[\gamma^2 + (\omega + \Omega)^2][\gamma^2 + (\omega - \Omega)^2]}, \tag{14}$$

and Ohmic spectral densities

$$J_O^s(\omega) = \frac{\lambda^2}{\pi} \frac{\omega^s}{s!\omega_c^{s-1}} e^{-\frac{\omega}{\omega_c}}, \tag{15}$$

defining a very important class of environments entering in the study of many systems, such as microscopic models leading to a Lindblad master equation for an harmonic oscillator in a weakly coupled high temperature environment, or a particle undergoing quantum Brownian motion [18,54,55]. From now on frequencies will be in cm^{-1} and temperatures in Kelvin. We remark that, because of the relation (8), if two spectral densities differ only for the overall coupling constant λ, their mappings (i.e., all of the chain coefficients ω_n, κ_n) will be identical, with the exception of κ_0, namely the coupling strength between the system ant the first TEDOPA chain mode. We also observe that the chain coefficients $\omega_n, k_n, n \geq 1$ are independent of the specific system-environment interaction term, that is, they are independent of the choice of A_S, O_ω of Equation (1). As customary for chain mappings, in what follows, we will moreover impose a hard cutoff ω_{hc} to the considered spectral densities, thus limiting their support to the interval $[0, \omega_{hc}]$ for $T = 0$ and to the interval $[-\omega_{hc}, \omega_{hc}]$ for $T > 0$. The value of ω_{hc} is suitably chosen as to keep the neglected relative reorganization energy

$$\frac{\int_{\omega_{hc}}^\infty d\omega J(\omega)/\omega}{\int_0^\infty d\omega J(\omega)/\omega}$$

in the order of 10^{-4} for all the considered instances.

If the considered spectral density belongs to the Szegö class, the asymptotic relations

$$\omega_\infty = \lim_{n\to\infty} \omega_n = \frac{\omega_{max} + \omega_{min}}{2}$$
$$\kappa_\infty = \lim_{n\to\infty} \kappa_n = \frac{\omega_{max} - \omega_{min}}{4} \tag{16}$$

hold (see Theorem 47 of Woods et al. [46]). Clearly enough, in our setting ω_{max} (and, at finite temperature, ω_{min}) depends on the imposed hard-cutoff ω_{hc} so that both ω_∞ and κ_∞ are functions of ω_{hc}. For any suitably fixed ω_{hc}, however, the relations (16) allow for the simple heuristic estimation $L = 2\kappa_\infty t_{max}$ of the maximal distance travelled within the time t_{max} by an excitation initially located at the first TEDOPA chain site. For fixed time t_{max}, therefore, the effective environment is made up of L oscillators within the "light-cone". Interestingly enough, the width of such light cone depends only on the "artificially" imposed hard cutoff and, as long as the choice ω_{hc} is sensibly chosen, different choices of the hard-cutoffs do not impact on the reduced dynamics of the system. On the other side, different spectral densities with the same support will have the same asymptotic coefficients, and the differences in the reduced dynamics of the system will be due to a (typically quite small) finite number of modes, as we will see in the following sections.

3. Chain Dynamics

We start by analyzing the dynamics of a single excitation moving along the chain-mapped environment produced by the (T-)TEDOPA mapping. To this end, we can disregard the system and the interaction term H_I, or equivalently set $\kappa_0 = 0$, and restrict our attention to the single excitation sector of the TEDOPA-chain Hilbert space. The set $\{|k\rangle,\ k = 1, 2, ...\}$, where $|k\rangle$ indicates the Fock state $|n_1 = 0, \ldots, n_{k-1} = 0, n_k = 1, n_{k+1} = 0, \ldots\rangle$ with the single excitation located at the k-th chain site, is a basis for the considered single excitation subspace. In what follows we will assume that the excitation is initially located at site 1, namely the initial state is $|1\rangle$.

3.1. Lorentzian Spectrum

The Lorentzian spectral density (14) provides a paradigmatic example. For $\gamma/\Omega \ll 1$, such spectrum well approximates that of an environment made up of a single harmonic oscillator with frequency Ω and dissipating into the vacuum at rate γ [18,56]. In all the following examples a hard cutoff frequency $\omega_{hc} = 10\Omega$ has been enforced.

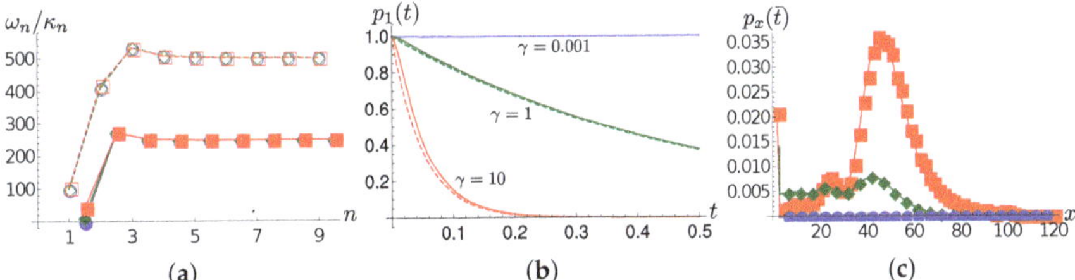

Figure 1. Lorentzian SD; in all frames $\Omega_S = 10$, $T = 0$; blue, green and red lines/marker refer respectively to $\gamma = 0.001$, $\gamma = 1$ and $\gamma = 10$. (**a**) The chain parameters ω_n (empty markers) and κ_n (filled markers) for $\gamma = 0.001$ (blue circles), $\gamma = 1$ (green diamonds) and $\gamma = 10$ (red squares); the couplings are shifted by 0.5 to the right to lie between n and $n+1$. (**b**) The population $p_1(t)$ of the first site as a function of time; the decay rates $\exp(-2\gamma)$ are shown as dashed lines as a guide to the eye. (**c**) The population of $p_x(\bar{t})$ at $\bar{t} = 0.2$ for $x = 1, 2, \ldots, 120$.

Frame (a) of Figure 1 shows the frequencies ω_n and couplings coefficients κ_n at $T = 0$ for $\Omega = 100$ and $\gamma = 0.001, 1, 10$ (see (14)). We first observe that, for all values of γ, the first and the second TEDOPA chain modes are equally far detuned. The main difference between the three selected cases lies in the coupling strength κ_1 between the same two modes, which is directly proportional to γ. The effect on the system dynamics is remarkable. As shown in Figure 1b the population of the first TEDOPA chain is well approximated by $p_1(t) = \exp(-2\gamma t)$, namely the decay rate of an harmonic oscillator damped into the vacuum at a rate γ. As frame (c) of the same figure shows, the portion of excitation that propagates beyond the first site propagates on the TEDOPA chain at a speed which is independent of γ: the chain coefficients are essentially equal to each other in the three cases for $n \geq 3$, and their value is determined by the hard cutoff frequency ω_{hc} through (16).

We turn now our attention to the finite temperature case.

As exemplified in Figure 2a, after the thermalization procedure [51] the thermalized spectral density (13) presents two peaks at $\pm\Omega$. The system will be thus effectively coupled to two damped modes, with temperature dependent coupling strength proportional to $1 + n_\beta(\Omega)$ resp. $n_\beta(\Omega)$.

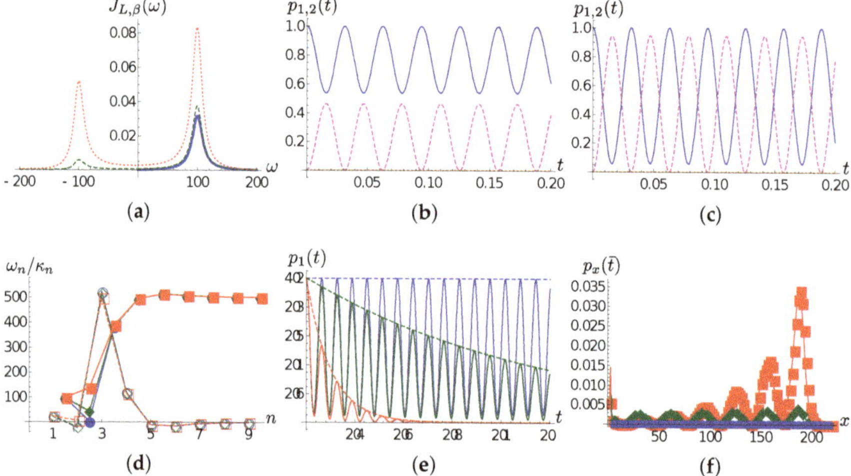

Figure 2. (**a**) The thermalized Lorentzian SD $J_{L,\beta}(\omega)$ (see Equations (13) and (14)) for $\Omega = 100, \gamma = 10$ at $T = 0$ (blue solid line), $T = 77$ (green dashed line) and $T = 300$ (red dotted line). In all the remaining frames $\Omega = 10, \gamma = 0.001$. (**b**) $T = 77$; the first (dashed blue line) and the second (magenta dashed line) TEDOPA chain site populations $p_{1,2}(t)$ as a function of time. (**c**) Same quantities and line styles as frame (**b**) at $T = 300$ (**d**–**f**): same quantities and styles as frames (**a**–**c**) of Figure 1 for $T = 300$.

It is thus not surprising that the chain dynamics for the case $\gamma = 0.001$ is essentially confined to the first two chain modes, as frames (b) and (c) of Figure 2 show. Indeed, the same plots suggest a clear relation between the temperature and the relative occupation of the modes: as T increase, the difference between the maxima of the populations of the first and the second TEDOPA chain sites decreases, and is expected to vanish as $T \to \infty$, that is, when the thermalized spectral density becomes symmetric with respect to the origin.

It is interesting to see that a mechanism very similar to the one discussed for the zero temperature case is at play also at finite temperature. Frame (d) of Figure 2 shows the chain coefficients for $\gamma = 0.001, 1$ and 10 at $T = 300$. This time the detuning between the first and the second TEDOPA chain sites is relatively small and the coupling between the two sites is independent of γ. This time it is the detuning between the second and the third chain site that is considerable, and the coupling κ_2 is monotone with γ. As shown in frame (e) of the same figure, the result is that the population of the first TEDOPA chain site presents damped beatings: the excitation moves forth and back between the first two chain sites, and percolates toward the right part of the chain at a rate $\exp(-2\gamma t)$. In the zero temperature case, the "escaped" population travels toward the right part of the chain at a speed which is independent of γ, and keeps trace of such beatings, as shown in Figure 2f, but this time the propagation speed is twice that of the zero temperature case because of the enlarged support $[-\omega_{hc}, \omega_{hc}]$ (see (16)).

In order to provide an insight on how the chain dynamics depends on the temperature, in Figure 3 we show the population of the first TEDOPA chain site for different values of T. As already observed, the decay rate and the frequency of the population oscillations are independent of T, which determines instead the amplitude of such oscillations. This leads us to the conclusion that the oscillation frequency must be determined by the parameter Ω, as expected.

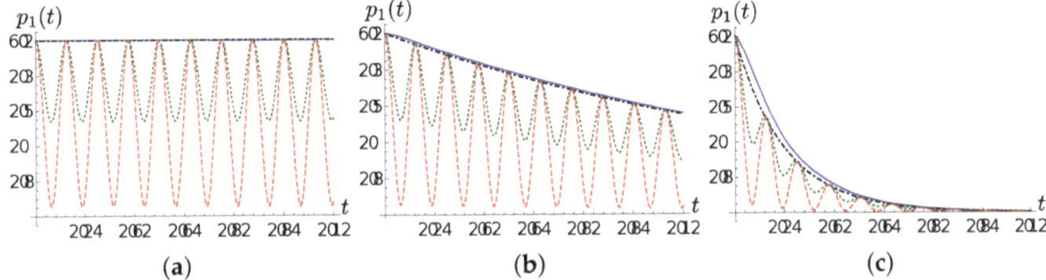

Figure 3. Lorentzian SD. The population $p_1(t)$ of the first TEDOPA chain as a function of time for $T = 0$ (blue solid line), $T = 77$ (green dotted line) and $T = 300$ (red dashed line) for **(a)** $\gamma = 0.001$, **(b)** $\gamma = 1$ and **(c)** $\gamma = 10$. In all plots $\exp(-2\gamma t)$ is show as a black dot-dashed line as a guide to the eye.

3.2. Ohmic Spectrum

We now consider spectral densities belonging to the Ohmic family, defined as in Equation 15. More in particular, we will study the chain dynamics on TEDOPA chains corresponding to the choice $s = 0.5, 1$ and 2, representative, respectively, of sub-Ohmic, Ohmic, and super-Ohmic spectral densities. In all the following examples we will set $\omega_c = 100$, and enforce a hard cutoff $\omega_{hc} = 10\omega_c$.

We start by the $T = 0$ case. Frames (a) and (b) of Figure 4 show Ohmic spectral densities for the selected values of s and the corresponding chain coefficients. The chain dynamics shows that an excitation leaves its initial location faster in the super-Ohmic case than in the Ohmic and sub-Ohmic case (see Figure 4c). This can be justified by the higher coupling coefficient and smaller detuning between the first sites of the TEDOPA chain in the $s = 2$ case with respect to the $s = 0.5, 1$ cases. Moreover, even if the front of the excitation wavepacket travels at the same speed in the three cases, the delocalization degree of the wavepacket is higher in the sub-Ohmic case, while it remains more "compact" in the super-Ohmic case, as examplified by the inset of Figure 4c, showing the TEDOPA chain site populations $p_x(t)$ at $t = 0.1$. Considered that the chain coefficients for $s = 0.5, 1, 2$ are very close to each other for $n \geq 4$, this difference is explained by the first chain coefficients. Roughly speaking, the higher coupling and smaller detuning between the first chain sites in the $s = 2$ case allows for more compact evolution of the wavepacket in the momentum space.

In the high-temperature regime $T = 300$, the main features of the chain dynamics are preserved, though with some differences. The decrease of population the first TEDOPA chain oscillator is still slower in the sub-Ohmic case; for the Ohmic SD, the first site population decay is similar to the $T = 0$ case, whereas for the super-Ohmic SD such decay is faster than in the zero temperature scenario (compare frames (c) and (f) of Figure 4). As already discussed before, this behaviour is mainly due to the detuning $|\omega_1 - \omega_2|$ and the coupling strength κ_1 between the first and second TEDOPA chain oscillators. Interestingly enough, for $s = 0.5$ part of the wavepacket remains localized at the first chain site, as shown in the inset of Figure 4f and, as in the T=0 case discussed above, the wavepacket is more delocalized in the sub-Ohmic case than in the super-Ohmic case, with the Ohmic case lying in between. As a last remark, we observe that, similarly to the finite temperature Lorentzian case, the propagation speed of the wavepacket is about twice as large as in the zero temperature case; as already discussed, this is due to the asymptotic relations (16).

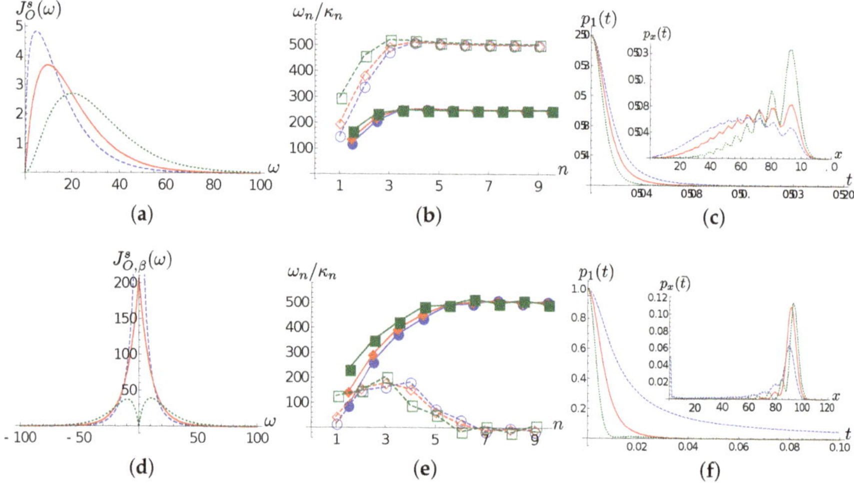

Figure 4. Ohmic SD. In all frames $\omega_c = 100$, and red markers/solid lines, blue markers/dashed lines, green markers/dotted lines correspond, respectively, to the Ohmic ($s = 1$), sub-Ohmic ($s = 0.5$) and super-Ohmic ($s = 2$) cases. (**a**) $T = 0$; the spectral density (15) for $s = 0.5, 1, 2$. (**b**) $T = 0$; the chain coefficients ω_n (empty markers), κ_n (filled markers). (**c**) The population of the first chain site as a function of time; in the inset, the populations $p_x(\bar{t})$ for $\bar{t} = 0.1$ as a function of x. (**d**) The thermalized SD $J^s_{O,\beta}(\omega)$ for $s = 0.5, 1, 2$ at $T = 300$. (**e**,**f**) Same quantities as frames (**b**,**c**) for $T = 300$.

Figure 5 provides more details. As we did for the Lorentzian SD case, we now inspect the dynamics of the first TEDOPA chain population for the three considered spectral densities at different temperatures. It clearly shows that, while for the Ohmic spectral density such population is only slightly affected by the value of T, the temperature has opposite effects on super- and sub-Ohmic SDs. As a matter of fact, whereas for the sub-Ohmic case, an increasing temperature leads to a slower decrease of the first site population, in the for $s = 2$ the first site empties at a rate which is directly proportional to the temperature. The snapshots on the populations $p_x(t)$ for $t = 0.02$ in the insets of frames (a)–(c) of the same figure, allows us to better appreciate the partial trapping at finite temperature of the wavepacket at the first TEDOPA chain site and the more pronounced spreading of the wavepacket in the $s = 0.5$ case.

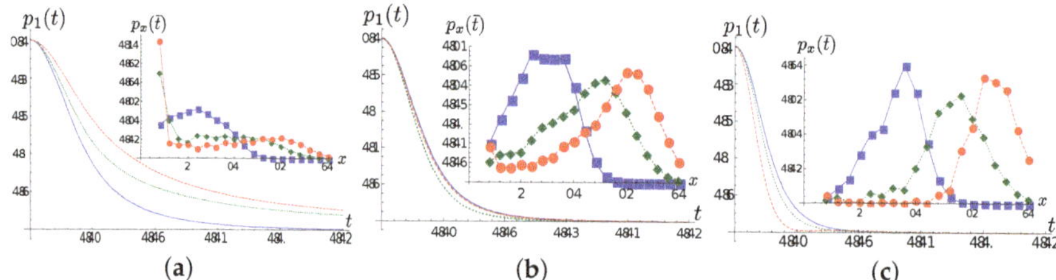

Figure 5. Ohmic SD. The population $p_1(t)$ of the first TEDOPA chain as a function of time for $T = 0$ (blue solid line), $T = 77$ (green dotted line) and $T = 300$ (red dashed line) for (**a**) $s = 0.5$, (**b**) $s = 1$ and (**c**) $s = 2$. In the inset of all frames, the population $p_x(\bar{t})$ at $\bar{t} = 0.02$.

4. Full Dynamics

So far we focused our analysis on the dynamics of a single excitation moving along TEDOPA chains. This allowed us to isolate the main features of such dynamics for representative spectral densities and to investigate the dependence of the kinematic properties of TEDOPA chains on the specific form of the SD and on the temperature. Clearly enough, the single excitation subspace we restricted ourselves to is not suited to describe the chain dynamics in the presence of a system interacting with the environment. As a matter of fact, the interaction with the system will dynamically inject in (and subtract from) the chain excitation, at a rate that depends, among other things, on the system-environment coupling strength.

In this section, therefore, we extend our analysis by considering a two-level system interacting with a bosonic environment described by either Lorentzian or Ohmic spectral densities. Given the spectral density, the spin-boson model is fully specified once the system and the system-environment interaction Hamiltonian are fixed. In what follows, we specialize the Hamiltonian (1) to

$$H_S = \Delta \sigma_x \tag{17}$$

$$A_S = \frac{1 + \sigma_z}{2} \tag{18}$$

$$O_\omega = X_\omega = (a_\omega + a_\omega^\dagger), \tag{19}$$

with σ_x, σ_z Pauli matrices, describing, for example, an homo-dimer interacting with a vibronic environment [57]. The resulting dynamics is therefore not a pure dephasing dynamics, and is representative of the class of physical systems for which numerically exact approaches are required. Considered that the interaction term does not change the system's populations but affects only its coherences, we will initialize the system to the state $|+\rangle = 1/\sqrt{2}(1,1)^T$, namely the eigenstate of σ_x belonging to the eigenvalue $+1$, representative of the maximally coherent states in the σ_z basis. The initial state of the environment will be instead a thermal state (11) at temperature T. In the following examples we will set $\Delta = 70 \text{cm}^{-1}$, and tune the parameter λ of Equations (14) and (15) so that the system-TEDOPA chain coupling κ_0 (see (8)) is the same at $T = 0$ for all the considered spectral densities. More precisely, by definition, the k_0 coefficient of the Ohmic spectral density is independent of s so that, in the Ohmic cases, we set $\lambda = 1$; for Lorentzian spectral densities we set to $\lambda = 60$.

Before presenting our results it is important to remark that we are not so much interested in the reduced dynamics of the system, but rather on the TEDOPA chain dynamics in the presence of an interaction with the open system. In particular, we will try to understand which of the features discussed in the preceding section persist in the presence of an interaction with the system. To this end we will use the average occupation number

$$n_k(t) = \text{Tr}(b_k^\dagger b_k \rho_C(t)) \tag{20}$$

of the k-th chain oscillator where $\rho_C(t)$ is the system+chain state at time t determined via TEDOPA simulation.

We first discuss the chain dynamics for Lorentzian spectral densities. The $\gamma = 0.001$ case is still paradigmatic. At $T = 0$ only the first TEDOPA chain oscillator is essentially involved in the dynamics. By comparing the purple lines in frames (a) and (b) of Figure 6, we can clearly see the beatings between the system and the first TEDOPA chain site. For $T > 0$ a the second TEDOPA chain mode enters into play. The average occupation number $n_{1,2}(t)$ of the first two chain sites depend on the temperature. Interestingly enough, in the high ($T = 300$) temperature regime the both $n_1(t)$ and $n_2(t)$ present small and fast out of phase oscillations, imprinting on the system dynamics a much more erratic dynamics than the $T = 77$ environment, for which such oscillations are slower and almost in phase.

Figure 7 shows instead the system and chain dynamics for $\gamma = 10$. Analogously to the $\gamma = 0.001$ the average occupation of the first two TEDOPA chain sites is temperature dependent. The larger value of γ implies that, loosely speaking, more environmental modes are interacting with the system. While the first two sites are still the highest occupied ones, some excitations can percolate to the right part of the chain, as we already observed in the chain dynamics analysis of the previous section (see Figure 2f). Since the system-TEDOPA chain coupling is about the same for the two considered values of γ, it is such percolation responsible for the faster relaxation of the system.

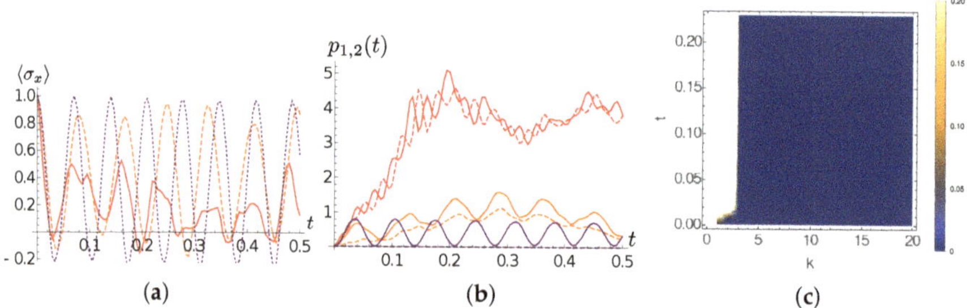

Figure 6. Lorentzian SD, full dynamics. $\gamma = 0.001$ (**a**) The expectation of σ_x as a function of time for $T = 0$ (purple dotted line) $T = 77$ (orange dashed line) and $T = 300$ (solid red line). (**b**) The average occupation number $p_{1,2}(t)$ of the first (solid lines) and the second (dashed line) TEDOPA chain sites for $T = 0$ (purple) $T = 77$ (orange) and $T = 300$ (red). (**c**) The average occupation number of the chain sites $k, k = 1, 2, \ldots, 20$ as a function of time for $T = 300$.

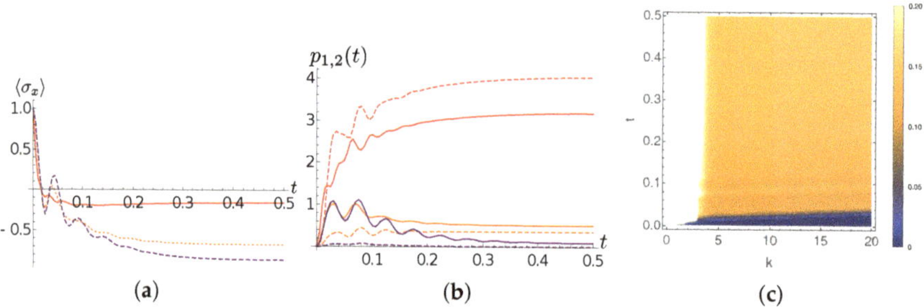

Figure 7. Lorentzian SD, full dynamics. Same quantities as in Figure 6 for $\gamma = 10$. (**a**) The expectation of σ_x as a function of time for $T = 0$ (purple dotted line) $T = 77$ (orange dashed line) and $T = 300$ (solid red line). (**b**) The average occupation number $p_{1,2}(t)$ of the first (solid lines) and the second (dashed line) TEDOPA chain sites for $T = 0$ (purple) $T = 77$ (orange) and $T = 300$ (red). (**c**) The average occupation number of the chain sites $k, k = 1, 2, \ldots, 20$ as a function of time for $T = 300$.

Now we turn our attention to Ohmic spectral densities. As it happens for the Lorentian case discussed above, the main features of the excitations dynamics presented in Section 3 provides a key to understanding the results. We observed (see Figure 4c,f) that an excitation located at the first chain site will leave its initial location more slowly in the sub-Ohmic case than in the Ohmic and super-Ohmic case. Moreover, the excitation wavepacket tends for $s = 0.5$ to be more spread over the chain than for $s = 2$, with the case $s = 1$ showing an intermediate behaviour. This features translate to the chain dynamics in the presence of an interaction with the system, as comparison between Figures 8–10 shows.

In more detail, we observe that at $T = 0$ the excitations leave the first chain sites almost ballisticaly for $s = 2$ (Figure 10b), whereas for $s = 0.5$ there is an accumulation of excitations in the very first part of the chain (Figure 8b). The diagonal fringes appearing in the sub-Ohmic (and less pronounced in the Ohmic) case at zero temperature are easily explained in therms of the (moving in time) population profile shown in the inset of Figure 4c. The inclination of the fringes, is instead related the the coupling coefficients beteween the TEDOPA chain oscillators that, as already pointed out, do not depend on *s* but only on the spectral density support.

At finite T the situation changes quite drastically. First of all we observe that for all the chosen values of *s* vertical fringes appear in frames (c) of Figures 8–10. Such vertical fringes can be associated to a the alternation of higher and lower average occupation number in nearest-neighbor sites, and allow to appreciate the onset of a stationary current when the state of the system gets close to its stationary state. A comparison between frames (a) of the same figures shows that in the sub-Ohmic the average occupation of the first TEDOPA chain sites is much higher than in the Ohmic and super-Ohmic cases. It must be noticed that, while the system-TEDOPA chain coupling κ_0 is equal for $T = 0$ for all values of *s*, at finite temperature such coupling is inversely proportional to *s*. The sub-Ohmic TEDOPA chain is therefore more strongly coupled to the system, and this justifies the faster system dynamics at short times.

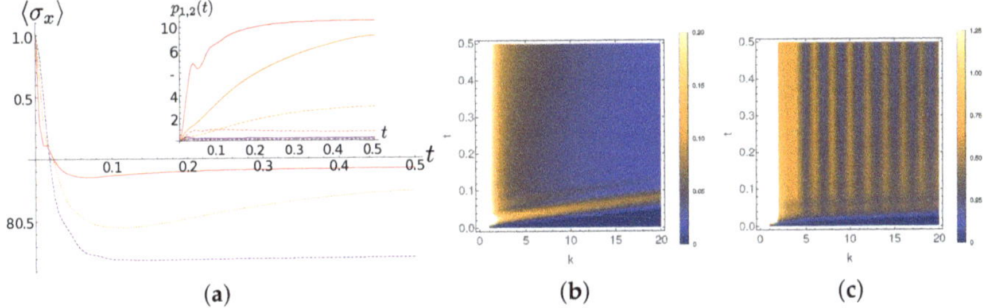

Figure 8. Sub-Ohmic SD ($s = 0.5$). (**a**) The expectation of σ_x at different temperatures as a function of time (same line styles as in Figure 6a); in the inset, the average occupation number of the first and the second TEDOPA chain oscillators (same line styles as in Figure 6b). (**b**) The average occupation number of the chain sites *k*, for $k = 1, 2, \ldots, 20$ as a function of time at $T = 0$. (**c**) Same quantities as in frame (**b**) for $T = 300$.

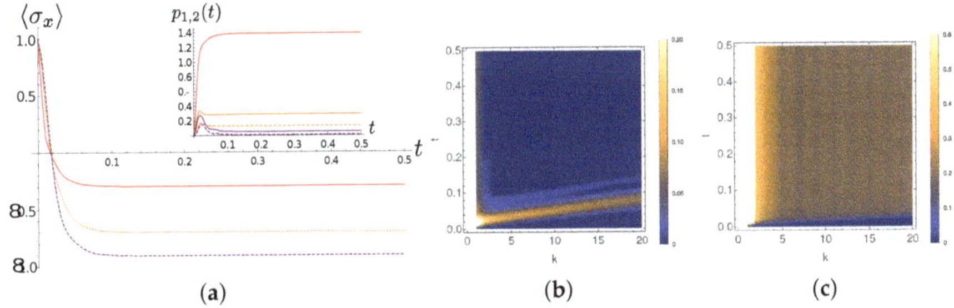

Figure 9. Ohmic SD ($s = 1$). Same quantities as in Figure 8. (**a**) The expectation of σ_x at different temperatures as a function of time (same line styles as in Figure 6a); in the inset, the average occupation number of the first and the second TEDOPA chain oscillators (same line styles as in Figure 6b). (**b**) The average occupation number of the chain sites *k*, for $k = 1, 2, \ldots, 20$ as a function of time at $T = 0$. (**c**) Same quantities as in frame (**b**) for $T = 300$.

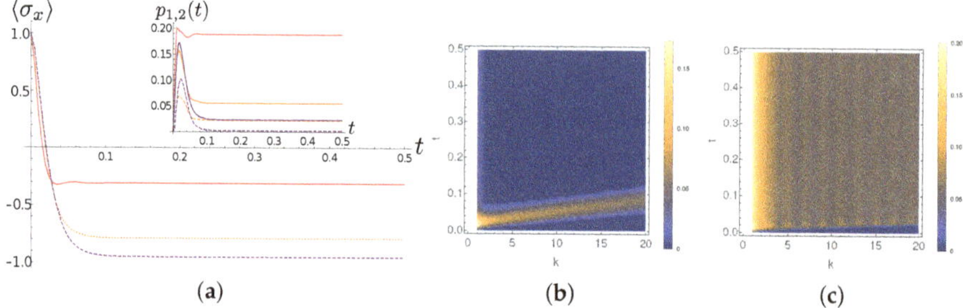

Figure 10. Super -Ohmic SD ($s = 2$). Same quantities as in Figure 8. (**a**) The expectation of σ_x at different temperatures as a function of time (same line styles as in Figure 6a); in the inset, the average occupation number of the first and the second TEDOPA chain oscillators (same line styles as in Figure 6b). (**b**) The average occupation number of the chain sites k, for $k = 1, 2, \ldots, 20$ as a function of time at $T = 0$. (**c**) Same quantities as in frame (**b**) for $T = 300$.

5. Conclusions and Outlook

While chain mapping has been recognized as a powerful tool for the efficient simulation of open quantum system dynamics, the subtle role of excitation dynamics on the determination of such reduced dynamics has never been investigated in detail. This work represents a first step in this direction. While the single excitation dynamics is unable to capture the full complexity of the evolution of TEDOPA chains put in interaction with the system, it provides a most useful key to understand such evolutions, as in the case of the Lorentzian spectral density we considered. It moreover provides a mean to sensibly set DMRG parameters, such as the chain truncation length and the local dimension of the chain oscillators: for super-Ohmic SDs, for example, the local dimension of the first TEDOPA chain oscillators must be set large enough as to host all the excitations that will accumulate in proximity of the system because of localization, while in the super-Ohmic case the local dimension of the first chain oscillators can be kept much smaller, since there is no signature of localization. While an analysis along the same lines for a specific spectral density was already presented [51], in this work we systematically compared and contrasted the features of the chain and full dynamics for a larger and very representative class of spectral densities. This allowed for example to shed light on the mechanisms allowing oscillators chain obtained by the T-TEDOPA procedure, and therefore starting from the vacuum state, to mimic an environment in the thermal state. For the Lorentzian case study such mechanism emerged quite clearly, and provided an key for the interpretation of the chain dynamics for SDs belonging to the Ohmic family.

We moreover observed that, while the asymptotic values of the TEDOPA coefficients determine the maximum distance reachable within a given time by an excitation initially located at the beginning of the TEDOPA chain, or light-cone, the features of a specific spectral density are typically determined by a very small number of coefficients. Indeed, as it happens in the $\gamma = 0.001$ Lorentzian SD case, the propagation of excitations in the light-cone can be hindered by an "effective" decoupling of the first sites of the chain from the remaining one. The analysis of the Ohmic SD instances, on the other side, showed that different (s-dependent) chain coefficients in the very fist part of the chain lead to quite different occupation probability profiles of the sites within the light-cone.

One of the, so far unexploited, advantages of chain mapping is the possibility of acquiring information on the state of the environment, something not meaningful when effective dynamics of Lindblad or Bloch-Redfield type are employed. While the number of chain modes perturbed by the interaction with the system is, in general, increasing with time, at any finite time it is in line of principle possible to make measurements on the oscillators in the light-cone. This could allow to understand, for example, which environmental modes are more involved in the dynamics and properly select the environmental reaction coordinates [58]. Moreover, in the presence of a fast convergence of the chain coefficients toward the asymptotic values, one expects a very small number of such coordinates. This represents a possible line of future research.

There are features of the TEDOPA chain evolution that remained quite obscure. For example, the fringes that appear in the Ohmic scenario at finite temperature are not present in the Lorentzian case. Considered that, as already observed, for $\gamma/\Omega \ll 1$ a Lorentzian environment can be assimilated to a damped harmonic oscillator undergoing a Lindblad-type dynamics, an therefore incoherently dissipating into an memoryless environment, one could read the lack of fringes as a signature of incoherent dynamics. A further analysis is therefore needed to better qualify the coherence dynamics in structured environments, and will be the focus of future work.

Funding: This research received no external funding.

Acknowledgments: The author acknowledges most useful discussions with Andrea Smirne during the development of this work and has been supported by UniMi through the "Sviluppo UniMi" project.

Conflicts of Interest: The author declares no conflict of interest.

References

1. May, V.; Kühn, P. *Charge and Energy Transfer Dynamics in Molecular Systems*; Wiley-VCH: Weinheim, Germany, 2004.
2. Plenio, M.B.; Huelga, S.F. Dephasing-assisted transport: quantum networks and biomolecules. *New J. Phys.* **2008**, *10*, 113019. [CrossRef]
3. Caruso, F.; Chin, A.W.; Datta, A.; Huelga, S.F.; Plenio, M.B. Highly efficient energy excitation transfer in light-harvesting complexes: The fundamental role of noise-assisted transport. *J. Chem. Phys.* **2009**, *131*, 105106. [CrossRef]
4. Chin, A.; Prior, J.; Rosenbach, R.; Caycedo-Soler, F.; Huelga, S.F.; Plenio, M.B. The role of non-equilibrium vibrational structures in electronic coherence and recoherence in pigment–protein complexes. *Nat. Phys.* **2013**, *9*, 113. [CrossRef]
5. Ferracin, D.; Mattioni, A.; Olivares, S.; Caycedo-Soler, F.; Tamascelli, D. Which-way interference within ringlike unit cells for efficient energy transfer. *Phys. Rev. A* **2019**, *99*, 062505. [CrossRef]
6. Anderson, P.W. Absence of diffusion in certain random lattices. *Phys. Rev.* **1958**, *109*, 1492–1505. [CrossRef]
7. Topinka, M.A.; LeRoy, B.J.; Westervelt, R.M.; Shaw, S.E.J.; Fleischmann, R.; Heller, E.J.; Maranowski, K.D.; Gossard, A.C. Coherent branched flow in a two-dimensional electron gas. *Nature* **2001**, *410*, 183–186. [CrossRef]
8. Baeriswyl, D.; Degiorgi, L.E. *Strong Interactions in Low Dimensions*; Kluwer Acedemic Publishers: Dordrecht, The Netherlands, 2004.
9. Hartmann, T.; Keck, F.; Korsch, H.J.; Mossmann, S. Dynamics of Bloch oscillations. *New J. Phys.* **2004**, *6*, 2. [CrossRef]
10. Benenti, G.; Casati, G.; Prosen, T.; Rossini, D. Negative differential conductivity in far-from-equilibrium quantum spin chains. *Europhys. Lett.* **2009**, *85*, 37001. [CrossRef]
11. Corrielli, G.; Crespi, A.; Della Valle, G.; Longhi, S.; Osellame, R. Fractional Bloch oscillations in photonic lattices. *Nat. Comm.* **2013**, *4*, 1555. [CrossRef]
12. Anufriev, R.; Ramiere, A.; Maire, J.; Nomura, M. Heat guiding and focusing using ballistic phonon transport in phononic nanostructures. *Nat. Comm.* **2017**, *8*, 15505. [CrossRef]
13. Partanen, M.; Tan, K.Y.; Govenius, J.; Lake, R.E.; Mäkelä, M.K.; Tanttu, T.; Möttönen, M. Quantum-limited heat conduction over macroscopic distances. *Nat. Phys.* **2016**, *12*, 460–464. [CrossRef] [PubMed]
14. Feynman, R.P. Simulating physics with computers. *Int. J. Theo. Phys.* **1982**, *21*, 467–488. [CrossRef]
15. Childs, A.M.; Cleve, R.; Deotto, E.; Farhi, E.; Gutmann, S.; Spielman, D.A. Exponential Algorithmic Speedup by a Quantum Walk. In *Proceedings of the Thirty-Fifth Annual ACM Symposium on Theory of Computing, San Diego, California, 9–11 June 2003*; Association for Computing Machinery: New York, NY, USA, 2003; STOC '03, pp. 59–68.
16. de Falco, D.; Tamascelli, D. Noise-assisted quantum transport and computation. *J. Phys. A Math. Theor.* **2013**, *46*, 225301. [CrossRef]

17. Benedetti, C.; Rossi, M.A.C.; Paris, M.G.A. Continuous-time quantum walks on dynamical percolation graphs. *Europhys. Lett.* **2019**, *124*, 60001. [CrossRef]
18. Breuer, H.P.; Petruccione, F. *The Theroy of Open Quantum Systems*; Oxford University Press: Oxford, UK, 2002.
19. Weiss, U. *Quantum Dissipative Systems*; World Scientific: Singapore, 2012.
20. Tamascelli, D.; Segati, A.; Olivares, S. Dephasing assisted transport on a biomimetic ring structure. *Int. J. Quant. Inf.* **2017**, *15*, 1740006. [CrossRef]
21. Luczka, J. Spin in contact with thermostat: Exact reduced dynamics. *Physica A* **1990**, *167*, 919. [CrossRef]
22. Hu, B.; Paz, J.; Zhang, Y. Quantum Brownian motion in a general environment: Exact master equation with nonlocal dissipation and colored noise. *Phys. Rev. D* **1992**, *45*, 2843. [CrossRef]
23. Garraway, B. Nonperturbative decay of an atomic system in a cavity. *Phys. Rev. A* **1997**, *55*, 2290. [CrossRef]
24. Fisher, J.; Breuer, H.P. Correlated projection operator approach to non-Markovian dynamics in spin baths. *Phys. Rev. A* **2007**, *76*, 052119. [CrossRef]
25. Smirne, A.; Vacchini, B. Nakajima-Zwanzig versus time-convolutionless master equation for the non-Markovian dynamics of a two-level system. *Phys. Rev. A* **2010**, *82*, 022110. [CrossRef]
26. Diósi, L.; Ferialdi, L. General Non-Markovian structure of Gaussian Master and Stochastic Schrödinger Equations. *Phys. Rev. Lett.* **2014**, *113*, 200403. [CrossRef] [PubMed]
27. Ferialdi, L. Exact closed master equation for Gaussian non-Markovian dynamics. *Phys. Rev. Lett.* **2016**, *116*, 120402. [CrossRef] [PubMed]
28. Huelga, S.; Plenio, M. Vibrations, quanta and biology. *Cont. Phys.* **2013**, *54*, 181–207. [CrossRef]
29. Rivas, A.; Huelga, S.F.; Plenio, M.B. Quantum non-Markovianity: characterization, quantification and detection. *Rep. Prog. Phys* **2014**, *77*, 094001. [CrossRef] [PubMed]
30. Breuer, H.P.; Laine, E.M.; Piilo, J.; Vacchini, B. Non-Markovian dynamics in open quantum systems. *Rev. Mod. Phys.* **2016**, *88*, 021002. [CrossRef]
31. De Vega, I.; Alonso, D. Dynamics of non-Markovian open quantum systems. *Rev. Mod. Phys* **2017**, *89*, 015001. [CrossRef]
32. Tanimura, Y.; Kubo, R. Two-Time Correlation Functions of a System Coupled to a Heat Bath with a Gaussian-Markoffian Interaction. *J. Phys. Soc. Jpn.* **1989**, *58*, 1199–1206. [CrossRef]
33. Ishizaki, A.; Fleming, G.R. Unified treatment of quantum coherent and incoherent hopping dynamics in electronic energy transfer: Reduced hierarchy equation approach. *J. Chem. Phys.* **1999**, *130*, 234111. [CrossRef]
34. Tanimura, Y. Stochastic Liouville, Langevin, Fokker–Planck, and Master Equation Approaches to Quantum Dissipative Systems. *J. Phys. Soc. Jpn.* **2006**, *75*, 082001. [CrossRef]
35. Feynman, R.P. Space-Time Approach to Non-Relativistic Quantum Mechanics. *Rev. Mod. Phys.* **1948**, *20*, 367–387. [CrossRef]
36. Makri, N.; Makarov, D.E. Tensor propagator for iterative quantum time evolution of reduced density matrices. I. Theory. *J. Chem. Phys.* **1995**, *102*, 4600–4610. [CrossRef]
37. Nalbach, P.; Eckel, J.; Thorwart, M. Quantum coherent biomolecular energy transfer with spatially correlated fluctuations. *New J. Phys.* **2010**, *12*, 065043. [CrossRef]
38. Somoza, A.D.; Marty, O.; Lim, J.; Huelga, S.F.; Plenio, M.B. Dissipation-Assisted Matrix Product Factorization. *Phys. Rev. Lett.* **2019**, *123*, 100502. [CrossRef]
39. Mascherpa, F.; Smirne, A.; Somoza, A.D.; Fernández-Acebal, P.; Donadi, S.; Tamascelli, D.; Huelga, S.F.; Plenio, M.B. Optimized auxiliary oscillators for the simulation of general open quantum systems. *Phys. Rev. A* **2020**, *101*, 052108. [CrossRef]
40. Lambert, N.; Ahmed, S.; Cirio, M.; Nori, F. Modelling the ultra-strongly coupled spin-boson model with unphysical modes. *Nat. Comm.* **2019**, *10*, 3721. [CrossRef]
41. Prior, J.; Chin, A.W.; Huelga, S.F.; Plenio, M.B. Efficient Simulation of Strong System-Environment Interactions. *Phys. Rev. Lett.* **2010**, *105*, 050404. [CrossRef]
42. Chin, A.W.; Rivas, A.; Huelga, S.F.; Plenio, M.B. Exact mapping between system-reservoir quantum models and semi-infinite discrete chains using orthogonal polynomials. *J. Math. Phys.* **2010**, *51*, 092109. [CrossRef]
43. Prior, J.; de Vega, I.; Chin, A.W.; Huelga, S.F.; Plenio, M.B. Quantum dynamics in photonic crystals. *Phys. Rev. A* **2013**, *87*, 013428. [CrossRef]

44. Hughes, K.; Christ, C.; Burghardt, I. Effective-mode representation of non-Markovian dynamics: A hierarchical approximation of the spectral density. I. Application to single surface dynamics. *J. Chem. Phys.* **2009**, *131*, 024109. [CrossRef]

45. Martinazzo, R.; Vacchini, B.; Hughes, K.; Burghardt, I. Universal Markovian reduction of Brownian particle dynamics. *J. Chem. Phys.* **2011**, *134*, 011101. [CrossRef]

46. Woods, M.P.; Groux, R.; Chin, A.W.; Huelga, S.F.; Plenio, M.B. Mappings of open quantum systems onto chain representations and Markovian embeddings. *J. Math. Phys.* **2014**, *55*, 032101. [CrossRef]

47. Ferialdi, L.; Dürr, D. Progress towards an effective non-Markovian description of a system interacting with a bath. *Phys. Rev. A* **2015**, *91*, 042130. [CrossRef]

48. White, S.R. Density matrix formulation for quantum renormalization groups. *Phys. Rev. Lett.* **1992**, *69*, 2863–2866. [CrossRef]

49. Woods, M.; Cramer, M.; Plenio, M. Simulating Bosonic Baths with Error Bars. *Phys. Rev. Lett.* **2015**, *115*, 130401. [CrossRef] [PubMed]

50. Leggett, A.J.; Chakravarty, S.; Dorsey, A.T.; Fisher, M.P.A.; Garg, A.; Zwerger, W. Dynamics of the dissipative two-state system. *Rev. Mod. Phys.* **1987**, *59*, 1. [CrossRef]

51. Tamascelli, D.; Smirne, A.; Lim, J.; Huelga, S.F.; Plenio, M.B. Efficient Simulation of Finite-Temperature Open Quantum Systems. *Phys. Rev. Lett* **2019**, *123*, 090402. [CrossRef] [PubMed]

52. Gautschi, W. Algorithm 726: ORTHPOL–a Package of Routines for Generating Orthogonal Polynomials and Gauss-type Quadrature Rules. *ACM Trans. Math. Softw.* **1994**, *20*, 21–62. [CrossRef]

53. Gautschi, W. *Orthogonal Polynomials Computation and Approximation*; Oxford Science Publications: Oxford, UK, 2004.

54. Caldeira, A.O.; Leggett, A.J. Path integral approach to quantum Brownian motion. *Phys. A Stat. Mech. Appl.* **1983**, *12*, 587–616. [CrossRef]

55. Ford, G.W.; Lewis, J.T.; O'Connell, R.F. Independent oscillator model of a heat bath: Exact diagonalization of the Hamiltonian. *J. Stat. Phys.* **1988**, *53*, 439–455. [CrossRef]

56. Lemmer, A.; Cormick, C.; Tamascelli, D.; Schaetz, T.; Huelga, S.F.; Plenio, M.B. A trapped-ion simulator for spin-boson models with structured environments. *New J. Phys.* **2018**, *20*, 073002. [CrossRef]

57. Plenio, M.B.; Almeida, J.; Huelga, S.F. Origin of long-lived oscillations in 2-D spectra of a quantum vibronic model: Electronic versus vibrational coherence. *J. Chem. Phys.* **2013**, *139*, 235102. [CrossRef] [PubMed]

58. Nazir, A.; Schaller, G. The Reaction Coordinate Mapping in Quantum Thermodynamic. In *Thermodynamics in the Quantum Regime. Fundamental Theories in Physics, vol. 195*; Binder, F., Correa, L., Gogolin, C., Anders, J., Adesso, G., Eds.; Springer: Cham, Switzerland, 2018.

Publisher's Note: MDPI stays neutral with regard to jurisdictional claims in published maps and institutional affiliations.

Article

Transport Efficiency of Continuous-Time Quantum Walks on Graphs

Luca Razzoli [1,*], Matteo G. A. Paris [2,3] and Paolo Bordone [1,4,*]

[1] Dipartimento di Scienze Fisiche, Informatiche e Matematiche, Università di Modena e Reggio Emilia, I-41125 Modena, Italy
[2] Quantum Technology Lab, Dipartimento di Fisica Aldo Pontremoli, Università Degli Studi di Milano, I-20133 Milano, Italy; matteo.paris@fisica.unimi.it
[3] INFN, Sezione di Milano, I-20133 Milano, Italy
[4] Centro S3, CNR-Istituto di Nanoscienze, I-41125 Modena, Italy
* Correspondence: luca.razzoli@unimore.it (L.R.); paolo.bordone@unimore.it (P.B.)

Abstract: Continuous-time quantum walk describes the propagation of a quantum particle (or an excitation) evolving continuously in time on a graph. As such, it provides a natural framework for modeling transport processes, e.g., in light-harvesting systems. In particular, the transport properties strongly depend on the initial state and specific features of the graph under investigation. In this paper, we address the role of graph topology, and investigate the transport properties of graphs with different regularity, symmetry, and connectivity. We neglect disorder and decoherence, and assume a single trap vertex that is accountable for the loss processes. In particular, for each graph, we analytically determine the subspace of states having maximum transport efficiency. Our results provide a set of benchmarks for environment-assisted quantum transport, and suggest that connectivity is a poor indicator for transport efficiency. Indeed, we observe some specific correlations between transport efficiency and connectivity for certain graphs, but, in general, they are uncorrelated.

Keywords: transport on graph; quantum walk; transport efficiency; connectivity

Citation: Razzoli, L.; Paris, M.G.A.; Bordone, P. Transport Efficiency of Continuous-Time Quantum Walks on Graphs. *Entropy* **2021**, *23*, 85. https://doi.org/10.3390/e23010085

Received: 27 November 2020
Accepted: 7 January 2021
Published: 9 January 2021

Publisher's Note: MDPI stays neutral with regard to jurisdictional claims in published maps and institutional affiliations.

1. Introduction

A continuous-time quantum walk (CTQW) is the quantum mechanical counterpart of the continuous-time random walk. It describes the dynamics of a quantum particle that continuously evolves in time in a discrete space, e.g., on the vertices of a graph, obeying the Schrödinger equation [1,2]. The Hamiltonian describing a CTQW is usually the Laplacian matrix L, which encodes the topology of the graph and it plays the role of the kinetic energy of the walker. Experimentally [3], CTQWs can be implemented on nuclear-magnetic-resonance quantum computers [4], optical lattices of ultracold Rydberg atoms [5], quantum processors [6], and photonic chips [7]. The applications of CTQWs range from implementing fast and efficient quantum algorithms [8,9], e.g., for spatial search [10] and image segmentation [11], to implementing quantum logic gates by multi-particle CTQWs in one-dimension (1D) [12], from universal computation [13] to modeling and simulating quantum phenomena, e.g., state transfer [14–16], quantum transport, and for characterizing the behavior of many-body systems [17,18].

Indeed, modeling quantum transport processes by means of CTQWs is a well-established practice and an appropriate mathematical framework. Quantum transport has been investigated with this approach on restricted geometries [19], semi-regular spider-net graphs [20], Sierpinski fractals [21], and on large-scale sparse regular networks [22]. CTQWs have been used in order to model transport of nonclassical light in coupled waveguides [23], coherent exciton transport on hierarchical systems [24], small-world networks [25], Apollonian networks [26], and on an extended star graph [27], coherent transport on complex networks [28], and exciton transfer with trapping [29,30]. It is worth noting that CTQWs do not necessarily perform better than their classical counterparts, since the transport properties strongly depend on the graph, the initial state, and on the propagation direction under investigation [31]. A measure of the efficiency of quantum and classical transport on graphs by means of the density of states has been proposed in [32].

Biological systems are known to show quantum effects [33,34] and efficient transport processes. Hence, the great interest in also studying CTWQs to model, e.g., exciton transport on dendrimers [35], photosynthetic energy transfer [36], environment-assisted quantum transport [37], dephasing-assisted transport on quantum networks and biomolecules [38], excitation transfer in light-harvesting systems [39,40], and its limits [41]. There also studies concerning disorder-assisted quantum transport on hypercubes and binary trees [42], because the latter can model a dendrimer-like structure for artificial light-harvesting systems [43,44].

Therefore, a full characterization of the transport properties on different structures is desired. Formally speaking, the CTQW Hamiltonian modeling transport processes shows similarities with the CTQW Hamiltonian adopted to study the spatial search. Both of them consist of the sum, with proper coefficients, of the Laplacian matrix, which is accountable for the motion of the walker on the graph, and the projector onto one or more specific vertices. This projector is the trapping Hamiltonian in transport problems and the oracle Hamiltonian in spatial search problems. The regularity, global symmetry, and connectivity of the graph have proved to be unnecessary for fast spatial search [45–47] by invoking certain graphs, e.g., complete bipartite graphs, strongly regular graphs, joined complete graphs, and a simplex of complete graphs, as counterexamples of these false beliefs. In this work, we address the transport by CTQW on the above mentioned graphs, which are different in terms of regularity, symmetry, and connectivity, and we assess the transport efficiency for initial states that are localized at a vertex and for an initial superposition of two vertices. Our focus is on the role of connectivity, if any. Indeed, regularity and global symmetry are not required for efficient transport, because removing some edges in the complete graph and the hypercube, which are regular and highly symmetric graphs, has been shown to improve the transport efficiency [48].

The paper is organized, as follows. In Section 2, we introduce CTQWs on a graph. In Section 3, we review the dimensionality reduction method to analyze CTQW problems [48], according to which we obtain a reduced model of the Hamiltonian encoding the problem that is considered and the reduced Hamiltonian still fully describes the dynamics that are relevant to the problem. In Section 4, we define the Hamiltonian modeling the transport on graphs and the transport efficiency as a figure of merit to measure the transport properties of the system. For each graph considered, we provide the reduced Hamiltonian and compute the transport efficiency for different initial states. In Section 5, we assess different measures of connectivity in order to characterize each graph considered. Finally, we present our conclusions in Section 6. In Appendix A, we report and refine the proof of the equality of the two subspaces that are required for computing the transport efficiency. In Appendix B, we determine the basis states spanning such a subspace for each graph considered.

2. Continuous-Time Quantum Walks

A graph is a pair $G = (V, E)$, where V denotes the non-empty set of vertices and E the set of edges. The order of the graph is the number of vertices, $|V| = N$. We define the adjacency matrix

$$A_{jk} = \begin{cases} 1 & \text{if } (j,k) \in E, \\ 0 & \text{otherwise,} \end{cases} \tag{1}$$

which describes the connectivity of G, and D the diagonal degree matrix with $D_{jj} = \deg(j)$, the degree of vertex j. In terms of these matrices, we introduce the graph Laplacian $L = D - A$, which is the matrix representation of the graph. According to this definition, L is positive semidefinite and singular.

The CTQW is the propagation of a quantum particle with kinetic energy when confined to a discrete space, e.g., a graph. The CTQW on a graph G takes place on a N-dimensional Hilbert space $\mathcal{H} = \mathrm{span}(\{|v\rangle \mid v \in V\})$, and the kinetic energy term ($\hbar = 1$) $T = -\nabla^2/2m$ is replaced by $T = \gamma L$, where $\gamma \in \mathbb{R}^+$ is the hopping amplitude of the walk. The state of the walker obeys the Schrödinger equation

$$i\frac{d}{dt}|\psi(t)\rangle = H|\psi(t)\rangle, \tag{2}$$

with Hamiltonian $H = \gamma L$. Hence, a walker starting in the state $|\psi_0\rangle \in \mathcal{H}$ continuously evolves in time, according to

$$|\psi(t)\rangle = U(t)|\psi_0\rangle, \tag{3}$$

with $U(t) = \exp[-iHt]$ the unitary time-evolution operator. The probability to find the walker in a target vertex w is therefore $|\langle w| \exp[-iHt]|\psi_0\rangle|^2$.

3. Dimensionality Reduction Method

In most CTQW problems, the quantity of interest is the probability amplitude at a certain vertex of the graph. The graph encoding the problem to solve often contains symmetries that allow for us to simplify the problem, since the evolution of the system actually occurs in a subspace of the complete N-dimensional Hilbert space $\mathcal{H}$ that is spanned by the vertices of the graph. We can determine the minimal subspace that contains the vertex of interest and it is invariant under the unitary time evolution via the dimensionality reduction method for CTQW, as proposed by Novo et al. [48], which we briefly review in this section for completeness. Such a subspace, also known as a Krylov subspace [49], contains the vertex of interest and all powers of the Hamiltonian applied to it. The relevance and the power of this method is that the graph encoding a given problem can be mapped onto an equivalent weighted graph, whose order is lower than the order of the original graph and whose vertices are the basis states of the invariant subspace. The corresponding reduced Hamiltonian still fully describes the dynamics that are relevant to the considered problem.

The unitary evolution (3) can be expressed as

$$|\psi(t)\rangle = \sum_{k=0}^{\infty} \frac{(-it)^k}{k!} H^k|\psi_0\rangle, \tag{4}$$

so $|\psi(t)\rangle$ is contained in the subspace $\mathcal{I}(H,|\psi_0\rangle) = \mathrm{span}(\{H^k|\psi_0\rangle \mid k \in \mathbb{N}_0\})$. This subspace of $\mathcal{H}$ is invariant under the action of the Hamiltonian and, thus, also of the unitary evolution. Naturally, $\dim \mathcal{I}(H,|\psi_0\rangle) \leq \dim \mathcal{H} = N$, but, if the Hamiltonian is highly symmetrical, only a small number of powers of $H^k|\psi_0\rangle$ are linearly independent, so the dimension of $\mathcal{I}(H,|\psi_0\rangle)$ can be much smaller than N.

Let P be the projector onto $\mathcal{I}(H,|\psi_0\rangle)$, so we have that

$$U(t)|\psi_0\rangle = PU(t)P|\psi_0\rangle = \sum_{k=0}^{\infty} \frac{(-it)^k}{k!} (PHP)^k|\psi_0\rangle = e^{-iPHPt}|\psi_0\rangle = e^{-iH_{\mathrm{red}}t}|\psi_0\rangle, \tag{5}$$

where $H_{\mathrm{red}} = PHP$ is the reduced Hamiltonian, and we used the fact that $P^2 = P$ (projector), $P|\psi_0\rangle = |\psi_0\rangle$, and $PU(t)|\psi_0\rangle = U(t)|\psi_0\rangle$.

For any state $|\phi\rangle \in \mathcal{H}$, which we consider to be the solution of the CTQW problem, we have

$$\langle\phi|U(t)|\psi_0\rangle = \langle\phi|PPU(t)P|\psi_0\rangle = \langle\phi|Pe^{-iH_{\mathrm{red}}t}|\psi_0\rangle = \langle\phi_{\mathrm{red}}|e^{-iH_{\mathrm{red}}t}|\psi_0\rangle, \tag{6}$$

where, the reduced state, $|\phi_{\text{red}}\rangle = P|\phi\rangle$. Reasoning analogously with the projector P' onto the subspace $\mathcal{I}(H, |\phi\rangle)$, we obtain

$$\langle\phi|U(t)|\psi_0\rangle = \langle\phi|e^{-iH'_{\text{red}}t}|\psi_{0\,\text{red}}\rangle,\tag{7}$$

with $H'_{\text{red}} = P'HP'$ and $|\psi_{0\,\text{red}}\rangle = P'|\psi_0\rangle$.

An orthonormal basis of $\mathcal{I}(H, |\phi\rangle)$, as denoted by $\{|e_1\rangle, \ldots, |e_m\rangle\}$, can be iteratively obtained, as follows: the first basis state is $|e_1\rangle = |\phi\rangle$, then the successive ones are obtained by applying H on the current basis state and orthonormalizing with respect to the previous basis states. The procedure stops when we find the minimum m such that $H|e_m\rangle \in \text{span}(\{|e_1\rangle, \ldots, |e_m\rangle\})$. The reduced Hamiltonian, i.e., H written in the basis of the invariant subspace, has a tridiagonal form, so the original problem is mapped onto an equivalent problem that is governed by a tight-binding Hamiltonian of a line with m sites.

4. Quantum Transport

The CTQW on a graph $G(V, E)$ of N vertices provides a useful framework to model, e.g., the dynamics of a particle or a quasi-particle (excitation) in a network. The quantum walker moves under the Hamiltonian

$$H = \gamma L = \gamma \sum_{i \in V} \deg(i)|i\rangle\langle i| - \gamma \sum_{(i,j) \in E} (|i\rangle\langle j| + |j\rangle\langle i|),\tag{8}$$

which can be read as a tight-binding Hamiltonian with uniform nearest-neighbor couplings γ and on-site energies $\gamma \deg(i)$. In the following, we set the units such that $\gamma = \hbar = 1$, so hereafter time and energy will be dimensionless.

However, in general, an excitation does not stay forever in the system in which it was created. In biological light-harvesting systems, the excitation gets absorbed at the reaction center, where it is transformed into chemical energy. In such a scenario, the total probability of finding the excitation within the network is not conserved. We assume a graph in which the walker can only vanish at one vertex $w \in V$, known as *trap vertex* or *trap*. The component of the walker's wave function at the trap vertex is absorbed by the latter at a trapping rate $\kappa \in \mathbb{R}^+$ [28]. Therefore, to phenomenologically model such loss processes we have to change the Hamiltonian (8), so we introduce the trapping Hamiltonian

$$H_{\text{trap}} = -i\kappa|w\rangle\langle w|,\tag{9}$$

which is anti-hermitian. This leads to the desired non-unitary dynamics that are described by the total Hamiltonian

$$H = L - i\kappa|w\rangle\langle w|.\tag{10}$$

This Hamiltonian has the same structure as the Hamiltonian for the spatial search of a marked vertex w [10], i.e., it is the sum of the Laplacian matrix and the projector onto $|w\rangle$, with proper coefficients. For spatial search, the projector onto $|w\rangle$ plays the role of the oracle Hamiltonian and the search Hamiltonian is hermitian. For quantum transport, the projector onto $|w\rangle$, because of the pure imaginary constant, plays the role of the trapping Hamiltonian (9) and the transport Hamiltonian (10) is not hermitian.

The transport efficiency is a relevant measure for a quantum transport process [37], which can be defined as the integrated probability of trapping at the vertex w

$$\eta = 2\kappa \int_0^{+\infty} \langle w|\rho(t)|w\rangle\, dt = 1 - \text{Tr}\left[\lim_{t \to +\infty} \rho(t)\right],\tag{11}$$

where $2\kappa\langle w|\rho(t)|w\rangle dt$ is the probability that the walker is successfully absorbed at the trap within the time interval $[t, t + dt]$ and $\rho(t) = |\psi(t)\rangle\langle\psi(t)|$ is the density matrix of the walker. The second equality of Equation (11) is due to the following reason. The surviving total probability of finding the walker within the graph at time t is $\langle\psi(t)|\psi(t)\rangle = \text{Tr}[\rho(t)]$

and it is ≤ 1 because of the loss processes at the trap vertex. Because the transport efficiency is the integrated probability of trapping in the limit of infinite time, we can also assess the transport efficiency as the complement to 1 of the probability of surviving within the graph, which is the complementary event.

In this scenario, there is no disorder in the couplings or site energies of the Hamiltonian or decoherence during the transport. In this ideal regime computing the transport efficiency amounts to finding the overlap of the initial state with the subspace $\Lambda(H, |w\rangle)$ spanned by the eigenstates of the Hamiltonian $|\lambda_k\rangle$ having a non-zero overlap with the trap $|w\rangle$, as proved by Caruso et al. [40]. Indeed, the dynamics are such that the component of the initial state within the space Λ is absorbed by the trap, whereas the component outside this subspace, i.e., in $\bar{\Lambda} = \mathcal{H} \setminus \Lambda$, remains in the graph (see Figure 1). Let us expand the initial state on the basis of the eigenstates of the Hamiltonian

$$|\psi_0\rangle = \sum_{k=1}^{m} \langle \lambda_k | \psi_0 \rangle | \lambda_k \rangle + \sum_{k=m+1}^{N} \langle \lambda_k | \psi_0 \rangle | \lambda_k \rangle = |\psi_\Lambda\rangle + |\psi_{\bar{\Lambda}}\rangle , \tag{12}$$

where we assume the eigenstates form an orthonormal basis (in the case of degenerate energy levels, we consider the eigenstates after orthonormalization) and are ordered in such a way that $\Lambda = \mathrm{span}(\{|\lambda_k\rangle \mid 1 \leq k \leq m\})$ and $\bar{\Lambda} = \mathrm{span}(\{|\lambda_k\rangle \mid m+1 \leq k \leq N\})$. The components in $\bar{\Lambda}$ are not affected by the open-dynamics that act at the trap vertex w. The remaining components evolve in the subspace Λ that is defined by having a finite overlap with the trap and are therefore absorbed at the trap. In the limit of $t \to +\infty$ the net result is the following: the total survival probability of finding the walker in the graph is $\langle \psi_{\bar{\Lambda}} | \psi_{\bar{\Lambda}} \rangle \leq 1$, i.e., it is due to the part of the initial state expansion in $\bar{\Lambda}$; instead, the part of the initial state expansion in Λ is fully absorbed at the trap, and so $\eta = \langle \psi_\Lambda | \psi_\Lambda \rangle = \sum_{k=1}^{m} |\langle \lambda_k | \psi_0 \rangle|^2$. A further consequence of this is that, if the system is initially prepared in a state $|\psi_0\rangle \in \bar{\Lambda}$, then the walker will stay forever in the graph without reaching the trap ($\eta = 0$); if the system is initially prepared in a state $|\psi_0\rangle \in \Lambda$, then the walker will be completely absorbed by the trap ($\eta = 1$).

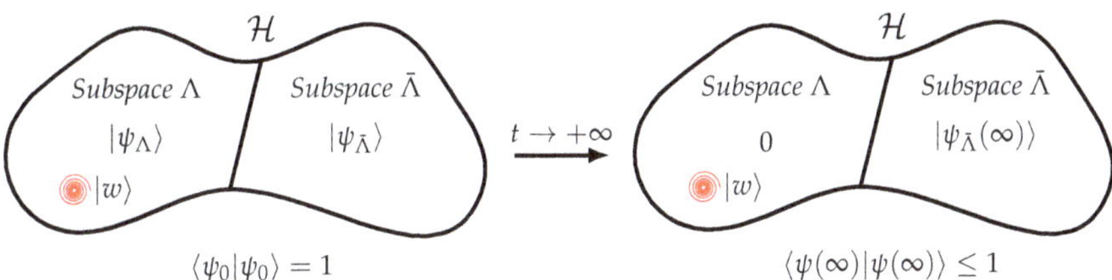

Figure 1. The quantum walker is in the initial state $|\psi_0\rangle$ (12) and it has components in $\Lambda(H, |w\rangle)$, the subspace spanned by the eigenstates of the Hamiltonian having a non-zero overlap with the trap $|w\rangle$, and in $\bar{\Lambda} = \mathcal{H} \setminus \Lambda$, the complement of Λ in the complete Hilbert space $\mathcal{H}$. In the limit of $t \to +\infty$, the dynamics are such that the component having non-zero overlap with the trap is fully absorbed by the trap, i.e., $|\psi_{\bar{\Lambda}}(\infty)\rangle = 0$, whereas the component in $\bar{\Lambda}$ survives. The dynamics are not unitary and the total survival probability of finding the walker within the graph is not conserved, i.e., $\langle \psi(\infty) | \psi(\infty) \rangle \leq 1$.

If, on the one hand, this analytical technique allows for one to compute the transport efficiency without solving dynamical equations, on the other hand diagonalizing the Hamiltonian might still be a hard task. The dimensionality reduction method in Section 3 allows for one to avoid diagonalizing the Hamiltonian, since it can be proved that $\Lambda(H, |w\rangle) = \mathcal{I}(H, |w\rangle)$ (see Appendix A). Hence, we compute the transport efficiency as

$$\eta = \sum_{k=1}^{m} |\langle e_k | \psi_0 \rangle|^2 , \tag{13}$$

i.e., as the overlap of the initial state $|\psi_0\rangle$ with the subspace $\mathcal{I}(H, |w\rangle) = \text{span}(\{|e_k\rangle \mid 1 \leq k \leq m\})$.

We consider as the initial state either a state localized at a vertex, $|\psi_0\rangle = |v\rangle$, or a superposition of two vertices, $|\psi_0\rangle = (|v_1\rangle + e^{i\theta}|v_2\rangle)/\sqrt{2}$. The localized initial state is a paradigmatic choice to take into account the fact that an excitation is usually created locally in a system. We also considered a superposition in order to investigate possible effects of coherence. The transport efficiency for the superposition of two vertices

$$\eta_s = \frac{1}{2} \sum_{k=1}^{m} \left| \langle e_k|v_1\rangle + e^{i\theta}\langle e_k|v_2\rangle \right|^2 \tag{14}$$

can be easily assessed, in some cases, when knowing the transport efficiency η_1 and η_2 for an initial state localized at v_1 and v_2, respectively. If $|v_1\rangle$ and $|v_2\rangle$ have the same overlap with the basis states, i.e., $\langle e_k|v_1\rangle = \langle e_k|v_2\rangle$ for $1 \leq k \leq m$, then $\eta_1 = \eta_2 = \eta$, and we have

$$\eta_s(\theta) = \frac{1}{2}\left|1 + e^{i\theta}\right|^2 \eta = (1 + \cos\theta)\eta, \tag{15}$$

so $0 \leq \eta_s(\theta) \leq 2\eta$. Instead, if $|v_1\rangle$ and $|v_2\rangle$ have nonzero overlap with different basis states, i.e., $\langle e_k|v_1\rangle \neq 0$ for $1 \leq k \leq m_1$ and $\langle e_k|v_2\rangle \neq 0$ for $m_1 + 1 \leq k \leq m_2$, with $m_2 \leq m$, then we have

$$\eta_s = \frac{1}{2}(\eta_1 + \eta_2), \tag{16}$$

and it is does not depend on θ.

In the following sections, we study quantum transport on different graphs that are relevant in terms of symmetry, regularity, and connectivity. For each graph, we determine the basis of the subspace in which the system evolves, the reduced Hamiltonian (10), and the transport efficiency (13) for an initial state localized at a vertex or a superposition of two vertices that is not covered by Equation (15). To analytically deal with a graph, we will group together the vertices that identically evolve by symmetry [45–47,50]. We mean that such vertices behave identically under the action of the Hamiltonian, in the sense that they are equivalent upon the relabeling of vertices, as well as, e.g., all of the vertices in a complete graph are equivalent. This does not mean that the time evolution $|v_1(t)\rangle$ of an initial state localized at a vertex v_1 is exactly equal to the time evolution $|v_2(t)\rangle$ of another initial state localized at $v_2 \neq v_1$, but it means that these two time evolutions are the same upon exchanging the labels of the two vertices. Note that the Hamiltonian (10) acts on a generic vertex as the Laplacian, except for the trap vertex, which, thus, forms a subset of one element, itself. The equal superpositions of the vertices in each subset form a orthonormal basis for a subspace of the Hilbert space and the Hamiltonian written in such a basis still fully describes the evolution of the system. However, we point out that such basis spans a subspace which, in general, is not the subspace $\mathcal{I}(H, |w\rangle)$ we need to compute the transport efficiency. Nevertheless, this grouping of vertices provides a useful framework to analytically deal with the system and, for this reason, we will introduce it. Clearly, identically evolving vertices have the same transport properties. However, vertices that are not equivalent for the Hamiltonian can provide the same transport efficiency. For this reason, in the following, we will stress when this is the case.

4.1. Complete Bipartite Graph

The complete bipartite graph (CBG) $G(V_1, V_2, E)$ is a highly symmetrical structure, which, in general, is not regular. The CBG has two sets of vertices, V_1 and V_2, such that each vertex of V_1 is only connected to all of the vertices of V_2 and vice versa. The set of CBGs is usually denoted as K_{N_1,N_2}, where the orders of the two partitions $N_1 = |V_1|$ and $N_2 = |V_2|$ are such that $N_1 + N_2 = N$, with N the total number of vertices. The CBG is non-regular as long as $N_1 \neq N_2$ (see $K_{4,3}$ in Figure 2), and the star graph is a particular case of CBG with $N_1 = N - 1$ and $N_2 = 1$. Without a loss of generality, we assume the trap vertex $w \in V_1$.

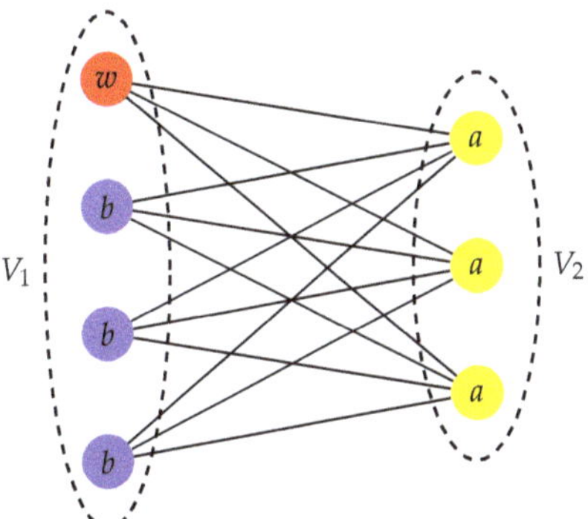

Figure 2. Complete bipartite graph $K_{4,3}$. The trap vertex $w \in V_1$ is colored red. Identically evolving vertices have the same transport properties and are identically colored and labeled.

The system evolves in a 3-dimensional subspace (see Appendix B.1) that is spanned by the orthonormal basis states

$$|e_1\rangle = |w\rangle, \quad |e_2\rangle = \frac{1}{\sqrt{N_2}} \sum_{i \in V_2} |i\rangle, \quad |e_3\rangle = \frac{1}{\sqrt{N_1 - 1}} \sum_{\substack{i \in V_1, \\ i \neq w}} |i\rangle. \tag{17}$$

This is also the basis that we would obtain by grouping together the identically evolving vertices in the subsets $V_a = V_2$ and $V_b = V_1 \setminus \{w\}$ (see Figure 2) [45]. In this subspace, the reduced Hamiltonian is

$$H = \begin{pmatrix} (1-\alpha)N - i\kappa & -\sqrt{(1-\alpha)N} & 0 \\ -\sqrt{(1-\alpha)N} & \alpha N & -\sqrt{(1-\alpha)(\alpha N - 1)N} \\ 0 & -\sqrt{(1-\alpha)(\alpha N - 1)N} & (1-\alpha)N \end{pmatrix}, \tag{18}$$

where $\alpha = N_1/N \in \mathbb{Q}^+$, $N_2 = (1-\alpha)N$, since $N_1 + N_2 = N$. Notice that, for G to be a CBG, α must satisfy the condition $1/N \leq \alpha \leq 1 - 1/N$.

If the initial state is localized at a vertex $v \neq w$, then the transport efficiency is

$$\eta = \begin{cases} \dfrac{1}{\alpha N - 1} & \text{if } v \in V_1 \,, \\[2ex] \dfrac{1}{(1-\alpha)N} & \text{if } v \in V_2 \,, \end{cases} \tag{19}$$

and we observe that

$$\eta_1 < \eta_2 \quad \Leftrightarrow \quad 2\alpha > 1 + \frac{1}{N}, \tag{20}$$

where $\eta_{1(2)} := \eta(v \in V_{1(2)})$. Instead, if the initial state is a superposition of two vertices, each of which belongs to a different partition, i.e., $v_1 \in V_1 \setminus \{w\}$ and $v_2 \in V_2$, then the transport efficiency

$$\eta_s = \frac{N-1}{2N(\alpha N - 1)(1-\alpha)} \tag{21}$$

follows from Equation (16), so clearly $\eta_{2(1)} \leq \eta_s \leq \eta_{1(2)}$, where the alternative depends on the condition (20). The transport efficiency depends on the parameters of the graph, N and α, as well as on the initial state (see Figure 3). Whether we consider an initial localized state or a superposition of two localized states, the asymptotic behavior is $\eta = O(1/N)$ if N_1 and N_2 are both sufficiently large.

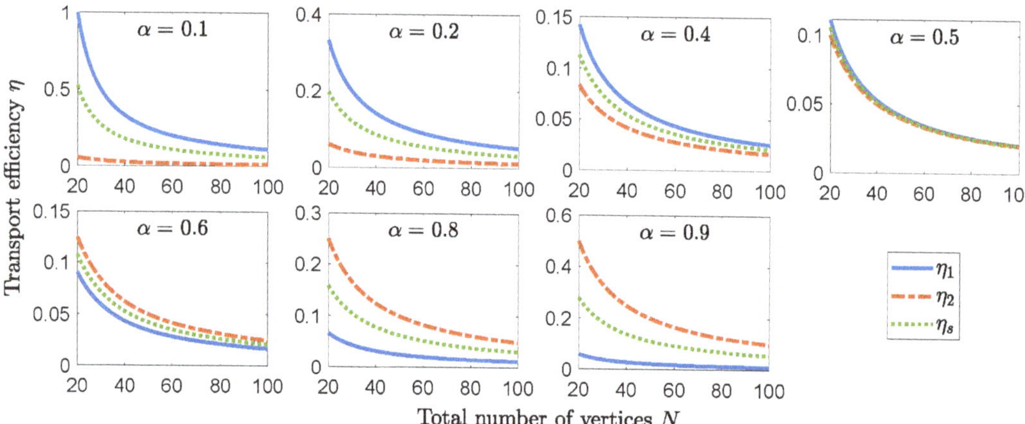

Figure 3. Transport efficiency η as a function of the order N of the complete bipartite graph for different values of $\alpha = N_1/N$, with $N_1 = |V_1|$, and different initial states. Transport efficiencies $\eta_{1(2)}$ (19) when the initial state is localized at a vertex in $V_{1(2)}$, and η_s (21) when the initial state is the superposition of two vertices, one in V_1 and the other in V_2. The trap vertex $w \in V_1$.

4.2. Strongly Regular Graph

A strongly regular graph (SRG) with parameters (N, k, λ, μ) is a graph with N vertices, not complete or edgeless, where each vertex is adjacent to k vertices; for each pair of adjacent vertices, there are λ vertices adjacent to both, and for each pair of nonadjacent vertices there are μ vertices that are adjacent to both [51,52]. If we consider the red vertex w in Figure 4, this means that there are k yellow adjacent vertices, and $N - k - 1$ blue vertices, all at distance 2. SRGs have a local symmetry, but most have no global symmetry [46]. The four parameters (N, k, λ, μ) are not independent and, for some parameters, there are no SRGs. One necessary, but not sufficient, condition is that the parameters satisfy

$$k(k - \lambda - 1) = (N - k - 1)\mu, \tag{22}$$

which can be proved by counting, in two wayy, the vertices at distance 0, 1, and 2 from a given vertex. Let us focus on the red vertex shown in Figure 4 and count the pairs of yellow and blue vertices that are adjacent to it. On the left-hand side of Equation (22), the red vertex has k neighbors, the yellow ones. Each yellow vertex has k neighbors, one of which is the red one and λ of which are other yellow vertices, so it is adjacent to $k - \lambda - 1$ blue vertices. Hence, the number of pairs of adjacent yellow and blue vertices is $k(k - \lambda - 1)$. On the right-hand side of Equation (22), we consider the blue vertices, which, by definition, are not adjacent to the red vertex. There are $N - k - 1$ blue vertices, since there are N total vertices in the graph, one of which is red and k of which are yellow. Each of the blue vertices is adjacent to μ yellow vertices, so there are $(N - k - 1)\mu$ pairs of yellow and blue vertices. The condition (22) comes from equating these expressions [46].

The system evolves in a 3-dimensional subspace (see Appendix B.2) spanned by the orthonormal basis states

$$|e_1\rangle = |w\rangle, \quad |e_2\rangle = \frac{1}{\sqrt{k}} \sum_{(i,w) \in E} |i\rangle, \quad |e_3\rangle = \frac{1}{\sqrt{N - k - 1}} \sum_{(i,w) \notin E} |i\rangle. \tag{23}$$

This is also the basis that we would obtain by grouping together the identically evolving vertices in the subsets $V_a = \{i \mid (i, w) \in E\}$ and $V_b = \{i \mid (i, w) \notin E\}$ (see Figure 4) [46]. In this subspace, the reduced Hamiltonian is

$$H = \begin{pmatrix} k - i\kappa & -\sqrt{k} & 0 \\ -\sqrt{k} & k - \lambda & -\sqrt{\mu(k - \lambda - 1)} \\ 0 & -\sqrt{\mu(k - \lambda - 1)} & \mu \end{pmatrix}. \tag{24}$$

If the initial state is localized at a vertex $v \neq w$, then the transport efficiency is

$$\eta = \begin{cases} \dfrac{1}{k} & \text{if } (v, w) \in E, \\[2ex] \dfrac{1}{N - k - 1} & \text{if } (v, w) \notin E. \end{cases} \tag{25}$$

Instead, if the initial state is a superposition of two vertices one of which is adjacent to w and the other is not, i.e., $(v_1, w) \in E$ and $(v_2, w) \notin E$, then the transport efficiency

$$\eta_s = \frac{N - 1}{2k(N - k - 1)} \tag{26}$$

follows from Equation (16).

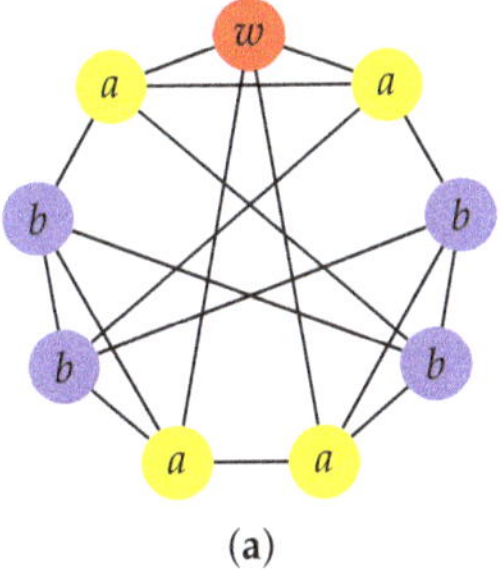
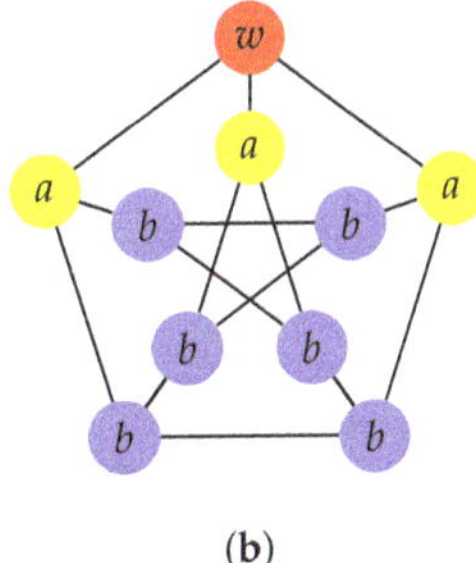

Figure 4. Two strongly regular graphs: (**a**) Paley graph with parameters $(9, 4, 1, 2)$ (parametrization (27) for $\mu = 2$); (**b**) Petersen graph with parameters $(10, 3, 0, 1)$. The trap vertex w is colored red. Identically evolving vertices have same transport properties and are identically colored and labeled.

A family of SRGs is the Paley graphs (see Figure 4a), which are parametrized by

$$(N, k, \lambda, \mu) = (4\mu + 1, 2\mu, \mu - 1, \mu) \tag{27}$$

where N must be a prime power (i.e., a prime or integer power of a prime [53]) such that $N \equiv 1 \pmod 4$. According to the parametrization (27), whether we consider an initial localized state or a superposition of two localized states, the transport efficiency on a Paley graph is $\eta = 1/2\mu$ (see Equations (25) and (26)), regardless of the fact that the vertices considered are adjacent or not to w.

4.3. Joined Complete Graphs

The transport efficiency on a complete graph, when the initial state is localized at a vertex $v \neq w$, is $\eta = 1/(N - 1)$ [40,48]. Here, we consider two complete graphs of $N/2$ vertices that are joined by a single edge (see Figure 5). The two vertices, b_1 and b_2, forming the "bridge" have degree $N/2$, whereas all of the others have degree $N/2 - 1$. We denote each complete graph by $K_{N/2}^{(k)} = (V_k, E_k)$, with $k = 1, 2$, where $|V_1| = |V_2| = N/2$. Therefore, the resulting joined graph is such that $V = V_1 \cup V_2$ and $E = E_1 \cup E_2 \cup \{(b_1, b_2)\}$.

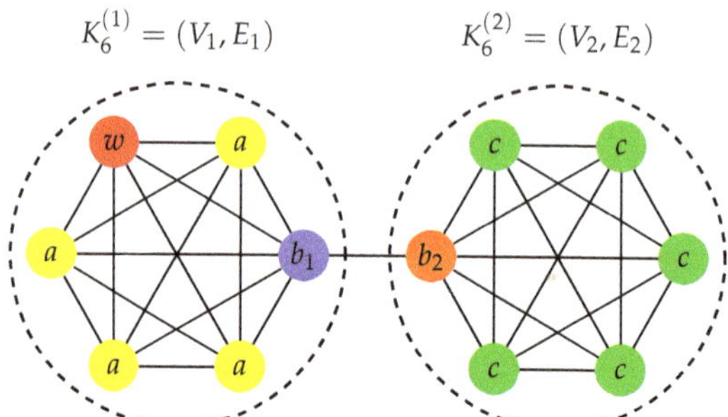

Figure 5. A graph with 12 vertices constructed by joining two complete graphs of 6 vertices by a single edge (b_1, b_2), the bridge. The trap vertex $w \in V_1$ is colored red. Identically evolving vertices have same transport properties and are identically colored and labeled. The vertices b_1 and b_2 show the same transport efficiency, even if they behave differently under the action of the Hamiltonian.

Grouping together the identically evolving vertices, we define the subsets $V_a = V_1 \setminus \{w, b_1\}$ and $V_c = V_2 \setminus \{b_2\}$ (see Figure 5). The system evolves in a 4-dimensional subspace (see Appendix B.3) that is spanned by the orthonormal basis states

$$|e_1\rangle = |w\rangle,$$

$$|e_2\rangle = \frac{1}{\sqrt{N/2-1}}\left(\sum_{i \in V_a} |i\rangle + |b_1\rangle\right),$$

$$|e_3\rangle = \frac{1}{\sqrt{(N-3)(N/2-1)}}\left[\sum_{i \in V_a} |i\rangle - (N/2-2)|b_1\rangle + (N/2-1)|b_2\rangle\right],$$

$$|e_4\rangle = \frac{1}{\sqrt{(N-3)[N(N/2-2)+1]}}\left[\sum_{i \in V_a} |i\rangle - (N/2-2)(|b_1\rangle + |b_2\rangle) - (N-3)\sum_{i \in V_c}|i\rangle\right]. \tag{28}$$

We point out that this basis spans a subspace of dimension 4, thus smaller than the 5-dimensional subspace spanned by the basis that is defined by grouping together the identically evolving vertices [47]. In the subspace that is spanned by the basis states $\{|e_1\rangle, \dots, |e_4\rangle\}$, the reduced Hamiltonian is

$$H = \begin{pmatrix} N/2-1-i\kappa & -\sqrt{N/2-1} & 0 & 0 \\ -\sqrt{N/2-1} & \frac{N}{N-2} & -\frac{\sqrt{N-3}}{N/2-1} & 0 \\ 0 & -\frac{\sqrt{N-3}}{N/2-1} & \frac{1}{N-3}\left(\frac{N^2}{2}-7+\frac{1}{N/2-1}\right) & \frac{\sqrt{(N/2-1)[N(N/2-2)+1]}}{N-3} \\ 0 & 0 & \frac{\sqrt{(N/2-1)[N(N/2-2)+1]}}{N-3} & \frac{N/2-1}{N-3} \end{pmatrix}. \tag{29}$$

If the initial state is localized at a vertex $v \neq w$, then the transport efficiency is

$$\eta = \begin{cases} \dfrac{2(N-1)}{N(N-4)+2} & \text{if } v \in V_a, \\[2ex] \dfrac{1}{2} + \dfrac{N-3}{N(N-4)+2} & \text{if } v \in \{b_1, b_2\}, \\[2ex] \dfrac{2(N-3)}{N(N-4)+2} & \text{if } v \in V_c. \end{cases} \tag{30}$$

Assuming that each complete graph has $N/2 \geq 3$ vertices, then $\eta_c < \eta_a \leq \eta_b$, where the subscript refers to an initial state localized at vertex in V_c, in V_a, and in the bridge $\{b_1, b_2\}$, respectively. Instead, if the initial state is a superposition of two vertices, then

$$
\eta_s(\theta) =
\begin{cases}
\dfrac{(N-2)[N+4(1+\cos\theta)]}{4[N(N-4)+2]} = \dfrac{1}{4} + O\left(\dfrac{1}{N}\right) & \text{if } v_1 \in V_a \wedge v_2 \in \{b_1, b_2\}\,, \\[2ex]
\dfrac{2(N-2-\cos\theta)}{N(N-4)+2} = \dfrac{2}{N} + O\left(\dfrac{1}{N^2}\right) & \text{if } v_1 \in V_a \wedge v_2 \in V_c\,, \\[2ex]
\dfrac{(N-2)[N-(N-4)\cos\theta]-4}{2[N(N-4)+2]} = \dfrac{1-\cos\theta}{2} + O\left(\dfrac{1}{N}\right) & \text{if } v_1 = b_1 \wedge v_2 = b_2\,, \\[2ex]
\dfrac{N(N+2)+4(N-4)\cos\theta-16}{4[N(N-4)+2]} = \dfrac{1}{4} + O\left(\dfrac{1}{N}\right) & \text{if } v_1 \in \{b_1, b_2\} \wedge v_2 \in V_c\,.
\end{cases}
\tag{31}
$$

We observe that, for the superposition of $v_1 \in V_a$ and $v_2 \in V_c$, the transport efficiency $\eta_s(\pi)$ is equal to η for an initial state that is localized at $v \in V_a$. For the superposition of b_1 and b_2, i.e., of the vertices of the bridge, we have $\eta_s(\pi) = 1$. This means that such a state belongs to $\mathcal{I}(H, |w\rangle)$, indeed

$$
\frac{1}{\sqrt{2}}(|b_1\rangle - |b_2\rangle) = \frac{1}{\sqrt{N-2}}(|e_2\rangle - \sqrt{N-3}|e_3\rangle)\,.
\tag{32}
$$

For an initial state localized at b_1 or b_2, we have the same transport efficiency η_b (30). However, the two vertices b_1 and b_2 have different overlap with the basis states $|e_k\rangle$, so the transport efficiency (31) for the superposition of them is not given by Equation (15).

4.4. Simplex of Complete Graphs

We call M-simplex of complete graphs what is formally known as the first-order truncated M-simplex lattice. The truncated M-simplex lattice is a generalization of the truncated tetrahedron lattice [54] and it is defined recursively. The graph of the zeroth order truncated M-simplex lattice is a complete graph of $M + 1$ vertices. The graph for the $(n + 1)$th order lattice is obtained by replacing each of the vertices of the nth order graph with a complete graph of M vertices. The truncated simplex lattice has been studied in various problems, e.g., in statistical models [55], self-avoiding random walks [56], and spatial search [47,57]. The M-simplex is, therefore, obtained by replacing each of the $M + 1$ vertices of a complete graph with a complete graph of M vertices (see Figure 6). Each of the new M vertices is connected to one of the edges coming to the original vertex. The graph is regular, vertex transitive, and there are $N = M(M + 1)$ total vertices.

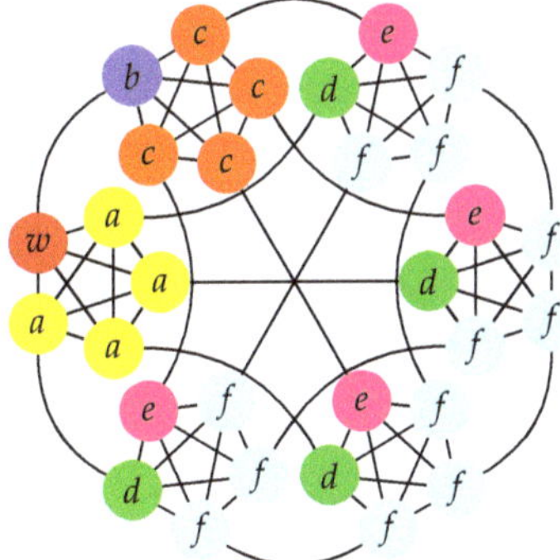

Figure 6. 5-simplex of complete graphs. The trap vertex w is colored red. Identically evolving vertices have same transport properties and are identically colored and labeled. The vertices in V_c and V_d show the same transport efficiency, even if they behave differently under the action of the Hamiltonian.

Grouping together the identically evolving vertices, we define the subsets V_a, V_c, V_d, V_e, and V_f (see Figure 6), having cardinality $|V_a| = |V_c| = |V_d| = |V_e| = M - 1$, and $|V_f| = (M-1)(M-2)$. The yellow vertices a are adjacent to w and belong to the same complete graph. The blue vertex b is adjacent to w, but it belongs to a different complete graph. The orange vertices c are adjacent to b and belong to the same complete graph. The green vertices d, even if, at distance 2 from w, like the vertices c, are adjacent to a, and so they form a different subset. The magenta vertices e are adjacent to c and belong to complete graphs other than the one the vertices c belong to. The cyan vertices f are adjacent to e and d. Independent of M, the system evolves in a 5-dimensional subspace (see Appendix B.4) that spanned by the orthonormal basis states

$$|e_1\rangle = |w\rangle,$$

$$|e_2\rangle = \frac{1}{\sqrt{M}}\left(\sum_{i\in V_a}|i\rangle + |b\rangle\right),$$

$$|e_3\rangle = \frac{\sqrt{M}}{\sqrt{(M-1)(M^2-2M+4)}}\left\{\frac{M-2}{M}\left[\sum_{i\in V_a}|i\rangle - (M-1)|b\rangle\right] + \sum_{i\in V_c\cup V_d}|i\rangle\right\},$$

$$|e_4\rangle = \frac{\sqrt{M^2-2M+4}}{\sqrt{(M-1)(M^3+2M^2-8M+16)}}\left\{\frac{2(M-2)}{M^2-2M+4}\left[\sum_{i\in V_a}|i\rangle - (M-1)|b\rangle\right]\right.$$

$$\left. -\frac{(M-2)^2}{M^2-2M+4}\sum_{i\in V_c\cup V_d}|i\rangle - 2\sum_{i\in V_e}|i\rangle - \sum_{i\in V_f}|i\rangle\right\},$$

$$|e_5\rangle = \frac{1}{M\sqrt{(M-1)(M-2)(M^3+2M^2-8M+16)}}\left\{-4(M-2)\left[\sum_{i\in V_a}|i\rangle - (M-1)|b\rangle\right]\right.$$

$$\left. +2(M-2)^2\sum_{i\in V_c\cup V_d}|i\rangle - M^2(M-2)\sum_{i\in V_e}|i\rangle + 2(M^2-2M+4)\sum_{i\in V_f}|i\rangle\right\}. \tag{33}$$

Note that, when the basis states include the vertices in V_c and V_d, they always involve the equal superposition of all the vertices in $V_c \cup V_d$. Thus, these vertices are equivalent for quantum transport, even if they behave differently under the action of the Hamiltonian. We point out that this basis spans a subspace of dimension 5, thus being smaller than the 7-dimensional subspace spanned by the basis that is defined by grouping together the identically evolving vertices [47,50]. In the subspace that is spanned by the basis states $\{|e_1\rangle, \ldots, |e_5\rangle\}$, the reduced Hamiltonian is a symmetric tridiagonal matrix with cumbersome elements, so we store the main diagonal and the superdiagonal, as follows

$$
\begin{pmatrix} H_{1,1} & H_{1,2} \\ \vdots & \vdots \\ H_{n,n} & H_{n,n+1} \\ \vdots & \vdots \\ H_{5,5} & * \end{pmatrix} =
\begin{pmatrix} M - i\kappa & -\sqrt{M} \\[4pt] \dfrac{3M-2}{M} & -\dfrac{\sqrt{(M-1)(M^2-2M+4)}}{M} \\[8pt] \dfrac{M^4-2M^3+4M^2-4M+8}{M(M^2-2M+4)} & \dfrac{\sqrt{M(M^3+2M^2-8M+16)}}{M^2-2M+4} \\[8pt] \dfrac{M(M^4-2M^3+20M^2-40M+64)}{(M^3+2M^2-8M+16)(M^2-2M+4)} & \dfrac{M(M+2)\sqrt{(M-2)(M^2-2M+4)}}{M^3+2M^2-8M+16} \\[8pt] \dfrac{(M+2)(M^3-4M+8)}{M^3+2M^2-8M+16} & * \end{pmatrix}, \tag{34}
$$

where the $*$ denotes the missing element, because its index exceeds the size of the matrix.

If the initial state is localized at a vertex $v \neq w$, then the transport efficiency is

$$\eta = \begin{cases} \dfrac{M^2 - 2}{M^2(M-1)} & \text{if } v \in V_a \,, \\[2ex] \dfrac{M^2 - 2M + 2}{M^2} & \text{if } v = b \,, \\[2ex] \dfrac{2}{M^2} & \text{if } v \in V_c \cup V_d \,, \\[2ex] \dfrac{1}{M-1} & \text{if } v \in V_e \,, \\[2ex] \dfrac{M^2 - 2M + 4}{M^2(M-1)(M-2)} & \text{if } v \in V_f \,. \end{cases} \tag{35}$$

Note that, for an initial state localized at b, which is the only vertex adjacent to w which does not belong to the complete graph of w (see Figure 6), we have $\eta_b \approx 1$ for large M. Instead, if the initial state is a superposition of two vertices, then

$$\eta_s(\theta) = \begin{cases} \dfrac{M(M^2 - 2M + 4) - 4 + 4(M-1)\cos\theta}{2M^2(M-1)} = \dfrac{1}{2} + O\left(\dfrac{1}{M}\right) & \text{if } v_1 \in V_a \wedge v_2 = b \,, \\[2ex] \dfrac{M^2 + 2M - 4 + 2(M-2)\cos\theta}{2M^2(M-1)} = \dfrac{1}{2M} + O\left(\dfrac{1}{M^2}\right) & \text{if } v_1 \in V_a \wedge v_2 \in V_c \cup V_d \,, \\[2ex] \dfrac{1}{M} + \dfrac{1}{M^2} & \text{if } v_1 \in V_a \wedge v_2 \in V_e \,, \\[2ex] \dfrac{M(M^2 - M - 4) + 8 - 4(M-2)\cos\theta}{2M^2(M-1)(M-2)} = \dfrac{1}{2M} + O\left(\dfrac{1}{M^2}\right) & \text{if } v_1 \in V_a \wedge v_2 \in V_f \,, \\[2ex] \dfrac{M^2 - 2M + 4 - 2(M-2)\cos\theta}{2M^2} = \dfrac{1}{2} + O\left(\dfrac{1}{M}\right) & \text{if } v_1 = b \wedge v_2 \in V_c \cup V_d \,, \\[2ex] \dfrac{1}{M^2} - \dfrac{1}{M} + \dfrac{M}{2(M-1)} = \dfrac{1}{2} + O\left(\dfrac{1}{M}\right) & \text{if } v_1 = b \wedge v_2 \in V_e \,, \\[2ex] \dfrac{M(M^3 - 5M^2 + 11M - 12) + 8}{2M^2(M-1)(M-2)} + \dfrac{2}{M^2}\cos\theta = \dfrac{1}{2} + O\left(\dfrac{1}{M}\right) & \text{if } v_1 = b \wedge v_2 \in V_f \,, \\[2ex] \dfrac{1}{M^2} + \dfrac{1}{2(M-1)} = \dfrac{1}{2M} + O\left(\dfrac{1}{M^2}\right) & \text{if } v_1 \in V_c \cup V_d \wedge v_2 \in V_e \,, \\[2ex] \dfrac{3M^2 - 8M + 8 + 2(M-2)^2\cos\theta}{2M^2(M-1)(M-2)} = \dfrac{3/2 + \cos\theta}{M^2} + O\left(\dfrac{1}{M^3}\right) & \text{if } v_1 \in V_c \cup V_d \wedge v_2 \in V_f \,, \\[2ex] \dfrac{1}{M^2} + \dfrac{1}{M} - \dfrac{1}{M-1} + \dfrac{1}{2(M-2)} = \dfrac{1}{2M} + O\left(\dfrac{1}{M^2}\right) & \text{if } v_1 \in V_e \wedge v_2 \in V_f \,. \end{cases} \tag{36}$$

Whenever the superposition of two vertices involves the vertex b, we have $\eta_s \approx 1/2$ for large M and, in particular, $\eta_s(\pi) = 1/2$ for $v_1 = b \wedge v_2 \in V_c \cup V_d$, independent of M (see Figure 7). Whenever the superposition involves a vertex in V_e, the transport efficiency does not depend on θ. Moreover, we observe that the equal superposition of the vertices in V_e belongs to $\mathcal{I}(H, |w\rangle)$, since

$$\frac{1}{\sqrt{M-1}} \sum_{i \in V_e} |i\rangle = -\frac{1}{\sqrt{M^3 + 2M^2 - 8M + 16}}\left(2\sqrt{M^2 - 2M + 4}|e_4\rangle + M\sqrt{M-2}|e_5\rangle\right), \tag{37}$$

and so this state provides $\eta = 1$.

In the M-simplex of complete graphs, the total number vertices is $N = M(M+1)$, so the asymptotic behavior of the transport efficiency must be understood, according to $M = O(\sqrt{N})$.

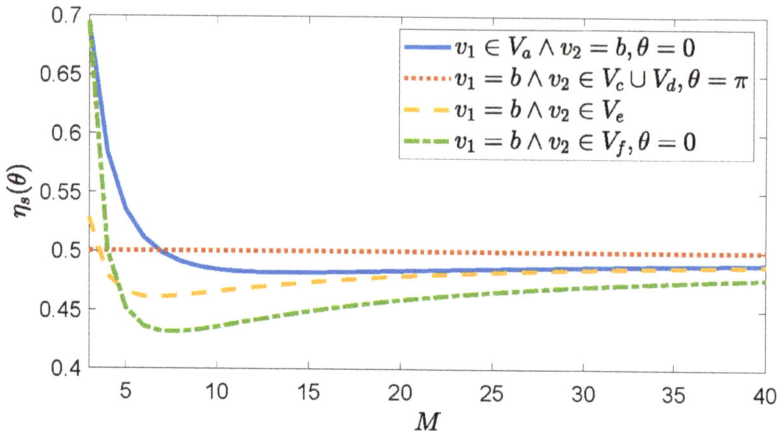

Figure 7. Transport efficiency $\eta_s(\theta)$ (36) as a function of M for different initial states $|\psi_0\rangle = (|v_1\rangle + e^{i\theta}|v_2\rangle)/\sqrt{2}$. M is the number of vertices in each of the $M+1$ complete graphs forming the M-simplex. The initial states are the possible equal superposition of two vertices, one of which is b.

5. Measures of Connectivity

The vertex connectivity $v(G)$ and edge connectivity $e(g)$ of a graph G are, respectively, the number of vertices or edges that we must remove to make G disconnected [58]. These are the two most common measures of graph connectivity, and

$$v(G) \leq e(G) \leq \delta(G), \tag{38}$$

i.e., both $v(G)$ and $e(G)$ are upper bounded by the minimum degree of the graph $\delta(G)$ [59].

Another measure follows from the Laplace spectrum of the graph. The second-smallest eigenvalue $a(G)$ of the Laplacian of a graph G with $N \geq 2$ vertices is the algebraic connectivity [60,61] and, to a certain extent, it is a good parameter to measure how well a graph is connected. In spectral graph theory it is well known, e.g., that a graph is connected if and only if its algebraic connectivity is different from zero. Indeed, the multiplicity of the Laplace eigenvalue zero of an undirected graph G is equal to the number of connected components of G [52]. For a complete graph, we know that $v(K_N) = e(K_N) = N-1$ and $a(K_N) = N$. Instead, for a noncomplete graph G, we have $a(G) \leq v(G)$, and so $a(G) \leq e(G)$ [58].

The results of the different measures of connectivity for each graph are shown in Table 1. Vertex, edge, and algebraic connectivities for the complete and the complete bipartite graphs are from [58]. The measures of connectivity for the M-simplex of complete graphs are from [47].

The vertex connectivity of a SRG is $v(G) = k$ [52] and the edge connectivity is $e(G) = k$. The latter follows from Equation (38), since $\delta(G) = k$, or using the fact that, if a graph has diameter 2, as the SRG has [62], then $e(G) = \delta(G)$ [59]. We need the Laplace spectrum in order to assess the algebraic connectivity. The eigenvalues of the adjacency matrix A are

$$\frac{1}{2}\left[\lambda - \mu \pm \sqrt{(\lambda - \mu)^2 + 4(k - \mu)}\right], \quad k, \tag{39}$$

and the scaling of them with N depends on the type of SRG. Indeed, SRGs can be classified into two types [51,59,62]. Type I graphs, for which $(N-1)(\mu - \lambda) = 2k$. This implies that $\lambda = \mu - 1$, $k = 2\mu$, and $N = 4\mu + 1$. They exist if and only if N is the sum of two squares. Examples include the Paley graphs (see parametrization (27)). Type II graphs, for which $(\mu - \lambda)^2 + 4(k - \mu)$ is a perfect square d^2, where d divides $(N-1)(\mu - \lambda) - 2k$, and the quotient is congruent to $N-1 \pmod 2$. Type I graphs are also type II graphs if and only if N is a square [51]. The Paley graph $(9, 4, 1, 2)$ is an example of this (see Figure 4a). Not all of the SRGs of type II are known, only certain parameter families, e.g., the Latin square graphs [51], and certain graphs, e.g., the Petersen graph (see Figure 4b), are. Hence, we consider the algebraic connectivity only for the SRGs of type I. According to the parametrization of the SRG of type I and to the fact that $D = kI$, the eigenvalues of $L = D - A$ are

$$0, \quad \frac{1}{2}(N \mp \sqrt{N}), \tag{40}$$

from which the algebraic connectivity is $a(G) = (N - \sqrt{N})/2$, since $\mu = (N-1)/4$ and $k = (N-1)/2$.

Table 1. The minimum degrees and vertex, edge, and algebraic connectivities of the graphs with N vertices that are considered in this work. For these graphs, the vertex and the edge connectivities are equal. Note that, in the M-simplex of complete graphs, $N = M(M+1)$.

Graph G	$\delta(G)$	$v(G) = e(G)$	$a(G)$
Complete K_N	$N - 1$	$N - 1$	N
Complete bipartite K_{N_1, N_2}	$\min(N_1, N_2)$	$\min(N_1, N_2)$	$\min(N_1, N_2)$
Strongly regular (Type I)	$(N-1)/2$	$(N-1)/2$	$(N - \sqrt{N})/2$
Joined complete $K_{N/2}$	$N/2 - 1$	1	$O(1/N)$
M-simplex	$M = O(\sqrt{N})$	$M = O(\sqrt{N})$	1

For the joined complete graphs we have $v(G) = e(G) = 1$, because of the bridge (see Figure 5) [63]. The Laplace spectrum is

$$0, \frac{N}{2}, \frac{1}{4}\left[N + 4 \pm \sqrt{N(N+8) - 16}\right], \tag{41}$$

from which the algebraic connectivity is $a(G) = [N + 4 - \sqrt{N(N+8) - 16}]/4$.

Subsequently, we assess whether connectivity of the graph may provide or not some bounds on the transport efficiency for an initial state localized at a vertex. First, we focus on the regular graphs considered in this work, for which $\delta(G) = v(G) = e(G)$, and this is equal to the degree. For a complete graph, we have $1/a(G) \leq \eta = 1/(N-1)$, and $1/(N-1)$ is also the reciprocal of the degree. For a SRG of type I, we have $\eta = 2/(N-1) \leq 1/a(G)$ for $\mu \geq 1$, and $2/(N-1)$ is also the reciprocal of the degree. Hence, from these two examples, we see that the reciprocal of the algebraic connectivity does not provide a common bound on η. For the M-simplex of complete graphs, we observe that $a(G) = 1$, from whose reciprocal we obtain the obvious upper bound $\eta \leq 1$. Note also that, in general, the transport efficiency for an initial state that is localized at vertex of a regular graph is not the reciprocal of the degree, as shown, e.g., by the transport efficiency on a general SRG (25) (degree k) and on the M-simplex (35) (degree M).

Now, we focus on the non-regular graphs. For the joined complete graphs, the reciprocal of the vertex and edge connectivity provides the obvious bound $\eta \leq 1$, whereas neither the reciprocal of $\delta(G)$ nor that of $a(G)$ provide a unique bound on η. Indeed, they are an upper or lower bound on η, depending on the initial state and the order of the graph (see Equation (30)). For the CBG, the vertex, edge, and algebraic connectivity is $\min(N_1, N_2)$ and its reciprocal is an upper or lower bound on the transport efficiency (19), depending on the geometry of the graph. Indeed, we have $\eta_1 \leq \eta_2 \leq 1/\min(N_1, N_2)$ for $\alpha > 1/2$, i.e., $N_1 > N_2$, and $1/\min(N_1, N_2) = \eta_2 \leq \eta_1$ for $\alpha \leq 1/2$, i.e., $N_1 \leq N_2$.

In conclusion, just by focusing on the transport efficiency for an initial state localized at a vertex, we observe that the connectivity is a poor indicator for the transport efficiency. First, because it does not provide any general lower or upper bound for estimating the transport efficiency, and transport efficiency and connectivity are generally uncorrelated (see Figure 8). Second, because transport efficiency strongly depends on the initial state, or, rather, on the overlap of this with the subspace spanned by the eigenstates of the

Hamiltonian having non-zero overlap with the trap vertex, as shown in Section 4. Note that, analogously, we have found no general correlation between the transport efficiency and normalized algebraic connectivity, which is the second-smallest eigenvalue of the normalized Laplacian matrix $\mathcal{L}$ of elements $\mathcal{L}_{jk} = L_{jk}/\sqrt{\deg(j)\deg(k)}$ [64].

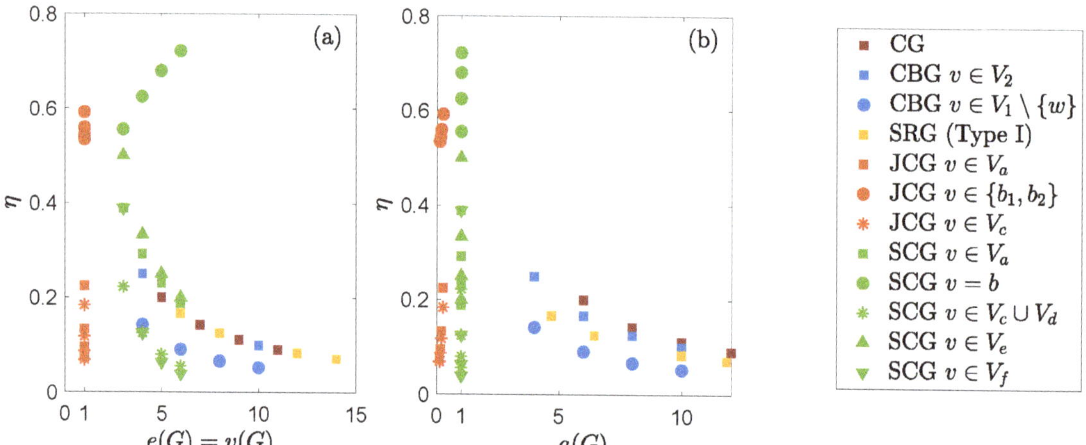

Figure 8. Scatter plot of the correlation between the transport efficiency η and (**a**) the edge or vertex connectivity, $e(G)$ and $v(G)$ respectively, or (**b**) the algebraic connectivity $a(G)$ (see also Table 1). Same color denotes results for the same graph: complete graph (CG, $N = 6, 8, 10, 12$), complete bipartite graph (CBG, $N = 12, 18, 24, 30, \alpha = 2/3$), strongly regular graphs of type I (SRG, $N = 13, 17, 25, 29$), joined complete graphs (JCG, $N = 12, 18, 24, 30$), and M-simplex of complete graphs (SCG, $M = 3, 4, 5, 6$). For a given a graph, different markers denote initial states localized at different vertices v. Note that, for the SRG of type I $\eta = 1/2\mu = 2/(N-1)$, independent of the fact that $(v, w) \in E$ or $(v, w) \notin E$. We observe some specific correlations between the transport efficiency and connectivity for a given graph, but, globally, among different graphs, transport efficiency and connectivity are uncorrelated.

6. Conclusions

In this work, we have addressed the coherent dynamics of transport processes on graphs in the framework of continuous-time quantum walks. We have considered graphs having different properties in terms of regularity, symmetry, and connectivity, and we have modeled the loss processes via the absorbing of the wavefunction component at a single trap vertex w. We have adopted the transport efficiency as a figure of merit in order to assess the transport properties of the system. In the ideal regime, as the one we have adopted, where there is no disorder or decoherence processes during the transport, the transport efficiency η can be computed as the overlap of the initial state with the subspace $\Lambda(H, |w\rangle)$ spanned by the eigenstates of the Hamiltonian having non-zero overlap with the trap vertex. According to the dimensionality reduction method, we have determined the orthonormal basis of such subspace with no need to diagonalize the Hamiltonian. Therefore, any initial state that is a linear combination of such basis states provides the maximum transport efficiency $\eta = 1$. We have considered, as the initial state, either a state localized at a vertex or a superposition of two vertices, and computed the corresponding transport efficiency. Overall, the most promising graph seems to be the M-simplex of complete graphs, since it allows for us to have a transport efficiency that is close to 1 for large M for an initially localized state. Transport with maximum efficiency is also possible on other graphs, if the walker is initially prepared in a suitable superposition state. However, the coherence of these preparations is likely to be degraded by noise, and the corresponding transport efficiency may be hard to be achieved in practice.

Our results suggest that connectivity of the graph is a poor indicator for the transport efficiency. Indeed, we observe some specific correlations between transport efficiency and connectivity for certain graphs, but in general they are uncorrelated. Moreover, transport efficiency depends on the overlap of the initial state with $\Lambda(H, |w\rangle)$, and the reciprocal of the measures of connectivity that we have assessed does not provide a general and consistent either lower or upper bound on η. However, the topology of the graph is encoded in the Laplacian matrix, which contributes to defining the Hamiltonian. Thus, connectivity somehow affects the transport properties of the system in the sense that it affects the Hamiltonian.

On the other hand, the transport efficiency is the integrated probability of trapping in the limit of infinite time, thus other figures of merit for the transport properties, such as the transfer time, which is the average time that is required by the walker to get absorbed at the trap, and the survival probability might highlight the role of the connectivity of the graph, if any. Moreover, the role of the trap needs to be further investigated, when considering more than one trap vertex, different trapping rates, and different trap location. Our analytical results are proposed as a reference for further studies on the transport properties of these systems and as a benchmark for studying environment-assisted quantum transport on such graphs. Indeed, our work paves the way for further investigation, including the analysis of more realistic systems in the presence of noise.

Author Contributions: Conceptualization, L.R., M.G.A.P. and P.B.; Methodology, L.R., M.G.A.P. and P.B.; Software, L.R.; Validation, L.R., M.G.A.P. and P.B.; Formal analysis, L.R., M.G.A.P. and P.B.; Investigation, L.R.; Writing—original draft preparation, L.R.; Writing—review and editing, L.R., M.G.A.P. and P.B.; Visualization, L.R.; Supervision, M.G.A.P. and P.B. All authors have read and agreed to the published version of the manuscript.

Funding: This research received no external funding.

Acknowledgments: Paolo Bordone and Matteo G. A. Paris are members of GNFM-INdAM.

Conflicts of Interest: The authors declare no conflict of interest.

Abbreviations

The following abbreviations are used in this manuscript:

CTQW	Continuous-time quantum walk
CBG	Complete bipartite graph
SRG	Strongly regular graph

Appendix A. Subspace of the Eigenstates of the Hamiltonian with Non-Zero Overlap with the Trap

In this appendix, we show that the subspace $\Lambda(H, |w\rangle)$ of the eigenstates of the Hamiltonian having nonzero overlap with the trap is equal to the subspace $\mathcal{I}(H, |w\rangle) = \text{span}(\{H^k|w\rangle \mid k \in \mathbb{N}_0\})$ introduced in Section 3. This proof is from the *Supplementary information* of [48]. We report it for sake of completeness and because we refine a key point, not addressed in the original proof, about the right and the left inverse of a matrix.

Let $\Lambda(H, |w\rangle) = \text{span}(\{|\lambda_1\rangle, \ldots, |\lambda_m\rangle\})$, where $H|\lambda_k\rangle = \lambda_k|\lambda_k\rangle$ and m is the minimum number of eigenstates of H having non-zero overlap with the trap, i.e., $\langle w|\lambda_k\rangle \neq 0$. In case of a degenerate eigenspace, more than one eigenstate belonging to it can have a non-zero overlap with $|w\rangle$, hence the need to find the minimum number m. The ambiguity is solved as follows. We choose the eigenstate from this degenerate eigenspace having the maximum overlap with $|w\rangle$, then we orthogonalize all the remaining eigenstates within such eigenspace with respect to it. After orthogonalizing, these eigenstates have zero overlap with $|w\rangle$ [40,48].

Let $\dim(\mathcal{I}(H, |w\rangle)) = m_1$, $\dim(\Lambda(H, |w\rangle)) = m_2$, and N the dimension of the complete Hilbert space. First, we prove that $\mathcal{I}(H, |w\rangle) \subseteq \Lambda(H, |w\rangle)$, i.e., that any state $H^i|w\rangle \in \mathcal{I}(H, |w\rangle)$ also belongs to $\Lambda(H, |w\rangle)$:

$$H^i|w\rangle = \sum_{k=1}^{N} \langle\lambda_k|w\rangle H^i|\lambda_k\rangle = \sum_{k=1}^{m_2} \langle\lambda_k|w\rangle H^i|\lambda_k\rangle = \sum_{k=1}^{m_2} \langle\lambda_k|w\rangle \lambda_k^i|\lambda_k\rangle, \tag{A1}$$

since $\langle\lambda_k|w\rangle = 0$ for $m_2 + 1 \leq k \leq N$. Any state $H^i|w\rangle$ can therefore be expressed as a linear combination of the eigenstates of the Hamiltonian having a non-zero overlap with the trap, so $H^i|w\rangle \in \Lambda(H, |w\rangle)\forall i \in \mathbb{N}_0$. Second, we prove that $\Lambda(H, |w\rangle) \subseteq \mathcal{I}(H, |w\rangle)$, i.e., that any state of $\Lambda(H, |w\rangle)$ can be expressed as a linear combination of the states of $\mathcal{I}(H, |w\rangle)$. We can write

$$|\lambda_j\rangle = \sum_{i=1}^{m_1} c_{ji} H^{i-1}|w\rangle = \sum_{k=1}^{m_2}\sum_{i=1}^{m_1} c_{ji}\lambda_k^{i-1}\langle\lambda_k|w\rangle|\lambda_k\rangle = \sum_{k=1}^{m_2}\sum_{i=1}^{m_1} c_{ji}M_{ik}|\lambda_k\rangle, \tag{A2}$$

with matrix element $M_{ik} = \lambda_k^{i-1}\langle\lambda_k|w\rangle$, provided that $\sum_{i=1}^{m_1} c_{ji}M_{ik} = \delta_{jk}$. In terms of matrices, this condition is $C_{m_2 \times m_1} M_{m_1 \times m_2} = I_{m_2 \times m_2}$, which means that C is the left inverse of M, i.e., $C = M_L^{-1}$. Analogously, rewriting Equation (A1) and then using the first equality of Equation (A2), we have

$$H^{j-1}|w\rangle = \sum_{i=1}^{m_2} \langle\lambda_i|w\rangle\lambda_i^{j-1}|\lambda_i\rangle = \sum_{i=1}^{m_2} M_{ji}|\lambda_i\rangle = \sum_{i=1}^{m_2}\sum_{k=1}^{m_1} M_{ji}c_{ik}H^{k-1}|w\rangle, \tag{A3}$$

provided that $\sum_{i=1}^{m_2} M_{ji}c_{ik} = \delta_{jk}$. In terms of matrices, this condition is $M_{m_1 \times m_2} C_{m_2 \times m_1} = I_{m_1 \times m_1}$, which means that C is the right inverse of M, i.e., $C = M_R^{-1}$. Therefore, M has a left and a right inverse, so M must be square, $m_1 = m_2 = m$, and $M_L^{-1} = M_R^{-1} = M^{-1} = C$ is unique [65]. The condition under which $\Lambda(H, |w\rangle) \subseteq \mathcal{I}(H, |w\rangle)$ is thus that M must be a $m \times m$ invertible matrix. The matrix M is invertible if $\det(M) \neq 0$. We define two $m \times m$ matrices, $V_{ij} = \lambda_j^{i-1}$ and the diagonal matrix $D_{ij} = \delta_{ij}\langle\lambda_j|w\rangle$, such that $M = VD$. Since $\langle\lambda_j|w\rangle = 0$ for $1 \leq j \leq m$, then $\det(V) \neq 0$. The matrix V is of the Vandermonde form, so $\det(V) = \prod_{1 \leq i < j \leq m}(\lambda_i - \lambda_j)$. This determinant is non-zero, since all of the states $|\lambda_k\rangle$, for $1 \leq k \leq m$, belong to different eigenspaces, so all the λ_k are different from each other. Hence, $\det(M) = \det(V)\det(D) \neq 0$, so M is always invertible and this condition ensures that $\Lambda(H, |w\rangle) \subseteq \mathcal{I}(H, |w\rangle)$. This concludes the proof that $\Lambda(H, |w\rangle) = \mathcal{I}(H, |w\rangle)$.

Appendix B. Basis of $\mathcal{I}(H, |w\rangle)$ for Each Graph

In this appendixm we analytically derive the orthonormal basis $\{|e_k\rangle\}$ spanning the subspace $\mathcal{I}(H, |w\rangle)$ for each graph considered. The first basis element is $|e_1\rangle = |w\rangle$, the trap vertex, and the k-th element $|e_k\rangle$ is obtained by orthonormalizing (O.N.) $H|e_{k-1}\rangle$ with respect to the subspace spanned by $\{|e_1\rangle, \ldots, |e_{k-1}\rangle\}$. The procedure stops when we find the minimum m, such that $H|e_m\rangle \in \mathrm{span}(\{|e_1\rangle, \ldots, |e_m\rangle\})$. The Hamiltonian (10) is the sum of the Laplacian matrix, generating the CTQW on the graph, and the trapping Hamiltonian (9), which projects onto the trap $|w\rangle$ with proper coefficient.

Appendix B.1. Complete Bipartite Graph

The Laplacian matrix of the CBG K_{N_1, N_2} is

$$L = N_2 \sum_{i \in V_1} |i\rangle\langle i| + N_1 \sum_{j \in V_2} |j\rangle\langle j| - \sum_{i \in V_1}\sum_{j \in V_2} (|i\rangle\langle j| + |j\rangle\langle i|), \tag{A4}$$

since $\deg(i \in V_1) = N_2$ and $\deg(j \in V_2) = N_1$ (see Figure 2). The basis states (17) are obtained, as follows:

$$H|e_1\rangle = (N_2 - i\kappa)|w\rangle - \sum_{j \in V_2} |j\rangle = (N_2 - i\kappa)|e_1\rangle - \sqrt{N_2}|e_2\rangle \xrightarrow{\text{O.N.}} |e_2\rangle, \tag{A5}$$

$$H|e_2\rangle = \frac{N_1}{\sqrt{N_2}} \sum_{j \in V_2} |j\rangle - \frac{1}{\sqrt{N_2}} \sum_{i \in V_1} \sum_{j \in V_2} |i\rangle = N_1|e_2\rangle - \sqrt{N_2} \sum_{\substack{i \in V_1, \\ i \neq w}} |i\rangle - \sqrt{N_2}|e_1\rangle$$

$$= N_1|e_2\rangle - \sqrt{N_2(N_1 - 1)}|e_3\rangle - \sqrt{N_2}|e_1\rangle \xrightarrow{\text{O.N.}} |e_3\rangle, \tag{A6}$$

$$H|e_3\rangle = \frac{N_2}{\sqrt{N_1 - 1}} \sum_{\substack{i \in V_1 \\ i \neq w}} |i\rangle - \frac{1}{\sqrt{N_1 - 1}} \sum_{\substack{i \in V_1 \\ i \neq w}} \sum_{j \in V_2} |j\rangle = N_2|e_3\rangle - \sqrt{N_2(N_1 - 1)}|e_2\rangle. \tag{A7}$$

In conclusion, any state $H^k|w\rangle \in \text{span}(\{|e_1\rangle, |e_2\rangle, |e_3\rangle\})\forall k \in \mathbb{N}_0$, thus the states (17) form an orthonormal basis for the subspace $\mathcal{I}(H, |w\rangle)$.

Appendix B.2. Strongly Regular Graph

The Laplacian matrix of the SRG with parameters (N, k, λ, μ) is

$$L = kI - \sum_{(j,i) \in E} |j\rangle\langle i|, \tag{A8}$$

where $I = \sum_{i \in V} |i\rangle\langle i|$ is the identity. Indeed, in a SRG each vertex has degree k, so the diagonal degree matrix is $D = kI$ (see Figure 4). The basis states (23) are obtained, as follows:

$$H|e_1\rangle = (k - i\kappa)|e_1\rangle - \sum_{(j,w) \in E} |j\rangle = (k - i\kappa)|e_1\rangle - \sqrt{k}|e_2\rangle \xrightarrow{\text{O.N.}} |e_2\rangle. \tag{A9}$$

A remark is due in order to address the computation of the next basis states. The diameter of a connected SRG G, i.e., the maximum distance between two vertices of G, is 2 [62]. This means that, given a vertex w, we can group all the other vertices in two subsets, as follows: the subset of the vertices at a distance 1 from w (adjacent); the subset of the vertices at a distance 2 from w (nonadjacent). Because of the structure of the SRG, where two (non)adjacent vertices have λ (μ) common adjacent vertices, in the following we face summations with repeated terms.

To determine the third basis state, we consider

$$H|e_2\rangle = k|e_2\rangle - \frac{1}{\sqrt{k}} \sum_{(i,w) \in E} \sum_{(j,i) \in E} |j\rangle = (k - \lambda)|e_2\rangle - \sqrt{k}|e_1\rangle - \sqrt{\mu(k - \lambda - 1)}|e_3\rangle \xrightarrow{\text{O.N.}} |e_3\rangle. \tag{A10}$$

To explain this, we have to focus on $\sum_{(i,w) \in E} \sum_{(j,i) \in E} |j\rangle$. The index of the first summation runs over the vertices i adjacent to w, whereas the index of the second summation runs over the vertices j adjacent to i. On the one hand, the vertex w is counted k times, because it has k adjacent vertices i, each of which, in turn, has $j = w$ among its adjacent vertices. On the other hand, the index of the second summation runs over the vertices adjacent and nonadjacent to w, because of the structure of the SRG. Each vertex j adjacent to w, i.e., $(j, w) \in E$, is connected to other λ vertices adjacent to w, so it is counted λ times. Each vertex j nonadjacent to w, i.e., $(j, w) \notin E$, is connected to μ vertices adjacent to w, so it is counted μ times. Thus, we have

$$\sum_{(i,w) \in E} \sum_{(j,i) \in E} |j\rangle = k|e_1\rangle + \lambda \sum_{(j,w) \in E} |j\rangle + \mu \sum_{(j,w) \notin E} |i\rangle = k|e_1\rangle + \lambda\sqrt{k}|e_2\rangle + \mu\sqrt{N - k - 1}|e_3\rangle. \tag{A11}$$

Accordingly, according to Equation (22), we can write $\mu\sqrt{(N-k-1)} = \sqrt{\mu k(k-\lambda-1)}$, from which Equation (A10) follows.

Subsequently, we consider

$$H|e_3\rangle = k|e_3\rangle - \frac{1}{\sqrt{N-k-1}} \sum_{(i,w)\notin E} \sum_{(j,i)\in E} |j\rangle = \mu|e_3\rangle - \sqrt{\mu(k-\lambda-1)}|e_2\rangle. \tag{A12}$$

Again, to explain this, we have to focus on the term $\sum_{(i,w)\notin E}\sum_{(j,i')\in E}|j\rangle$ in the second equality. The index of the first summation runs over the vertices i nonadjacent to w, whereas the index of the second summation runs over the vertices j adjacent to i. Each vertex j nonadjacent to w, i.e., $(j,w)\notin E$, is connected to other $k-\mu$ vertices nonadjacent to w, so it is counted $k-\mu$ times. Each vertex j adjacent to w, i.e., $(j,w)\in E$, is connected to $k-\lambda-1$ vertices nonadjacent to w, so it is counted $k-\lambda-1$ times. Thus, we have

$$\sum_{(i,w)\notin E} \sum_{(j,i)\in E} |j\rangle = (k-\lambda-1) \sum_{(i,w)\in E} |i\rangle + (k-\mu) \sum_{(i,w)\notin E} |i\rangle$$

$$= (k-\lambda-1)\sqrt{k}|e_2\rangle + (k-\mu)\sqrt{N-k-1}|e_3\rangle. \tag{A13}$$

So, according to Equation (22), we can write $(k-\lambda-1)\sqrt{k} = \sqrt{\mu(N-k-1)(k-\lambda-1)}$, from which Equation (A12) follows.

In conclusion, any state $H^k|w\rangle \in \mathrm{span}(\{|e_1\rangle, |e_2\rangle, |e_3\rangle\})\forall k \in \mathbb{N}_0$, thus the states (23) form an orthonormal basis for the subspace $\mathcal{I}(H, |w\rangle)$.

Appendix B.3. Joined Complete Graphs

The Laplacian matrix of the two complete graphs $K_{N/2}$ joined by a single edge (b_1, b_2) is

$$L = L_1 + L_2 + \underbrace{|b_1\rangle\langle b_1| + |b_2\rangle\langle b_2| - |b_1\rangle\langle b_2| - |b_2\rangle\langle b_1|}_{\text{bridge}}, \tag{A14}$$

where

$$L_k = \left(\frac{N}{2}-1\right) \sum_{i\in V_k} |i\rangle\langle i| - \sum_{(i,j)\in E_k} |i\rangle\langle j| \tag{A15}$$

is the Laplacian matrix of the complete graph $K_{N/2}^{(k)}$, with $k=1,2$. The bridge introduces the edge between the vertices b_1 and b_2 and correctly makes the degree of such vertices be $N/2$ (see Figure 5). Hence, $L|v\rangle = L_k|v\rangle$ for any vertex $v \in V_k \setminus \{b_k\}$. Instead, $L|b_k\rangle = (N/2)|b_k\rangle - \sum_{(i,b_k)\in E_k}|i\rangle - |b_{\bar{k}}\rangle$, where $\bar{k}$ is the complement of k in $\{1,2\}$.

Reasoning by symmetry, we introduce the subsets of the identically evolving vertices, i.e., the subsets containing the vertices that behave identically under the action of the Hamiltonian:

$$H|w\rangle = (N/2-1-i\kappa)|w\rangle - \sum_{i\in V_a} |i\rangle - |b_1\rangle, \tag{A16}$$

$$H\sum_{i\in V_a}|i\rangle = 2\sum_{i\in V_a}|i\rangle - (N/2-2)(|w\rangle+|b_1\rangle), \tag{A17}$$

$$H|b_1\rangle = N/2|b_1\rangle - \sum_{i\in V_a}|i\rangle - |w\rangle - |b_2\rangle, \tag{A18}$$

$$H|b_2\rangle = N/2|b_2\rangle - \sum_{i\in V_c}|i\rangle - |b_1\rangle, \tag{A19}$$

$$H\sum_{i\in V_c}|i\rangle = \sum_{i\in V_c}|i\rangle - (N/2-1)|b_2\rangle, \tag{A20}$$

where $V_a = V_1 \setminus \{w, b_1\}$ and $V_c = V_2 \setminus \{b_2\}$. Note that the results of H applied on the vertices b_1 or b_2 are different, and this is the reason why they form different subsets. According to these preliminary results, the basis states (28) are obtained, as follows:

$$H|e_1\rangle = (N/2 - 1 - i\kappa)|w\rangle - \sum_{i \in V_a} |i\rangle - |b_1\rangle \xrightarrow{\text{O.N.}} |e_2\rangle, \tag{A21}$$

$$H|e_2\rangle = \frac{1}{\sqrt{N/2-1}}\left[-(N/2-1)|w\rangle + \sum_{i \in V_a} |i\rangle + 2|b_1\rangle - |b_2\rangle \right] \xrightarrow{\text{O.N.}} |e_3\rangle, \tag{A22}$$

$$H|e_3\rangle = \frac{1}{\sqrt{(N-3)(N/2-1)}}\left[N/2 \sum_{i \in V_a} |i\rangle - (N^2/4 - 3)|b_1\rangle + (N^2/4 - 2)|b_2\rangle \right.$$

$$\left. -(N/2-1)\sum_{i \in V_c} |i\rangle \right] \xrightarrow{\text{O.N.}} |e_4\rangle, \tag{A23}$$

and it can be proved that

$$H|e_4\rangle = \frac{\sqrt{N/2-1}}{N-3}\left(\sqrt{N(N/2-2)+1}\,|e_3\rangle + \sqrt{N/2-1}\,|e_4\rangle \right). \tag{A24}$$

In conclusion, any state $H^k|w\rangle \in \text{span}(\{|e_1\rangle, \ldots, |e_4\rangle\}) \forall k \in \mathbb{N}_0$, thus the states (28) form an orthonormal basis for the subspace $\mathcal{I}(H, |w\rangle)$.

Appendix B.4. Simplex of Complete Graphs

The Laplacian matrix is defined as $L = D - A$. For a M-simplex of complete graphs the diagonal degree matrix is $D = MI$, since the graph is regular, and the adjacency matrix is $A = \sum_{m=1}^{M+1} A_{\text{intra}}^{(m)} + A_{\text{inter}}$, where

$$A_{\text{intra}}^{(m)} = \sum_{(i,j) \in E_m} |i^{(m)}\rangle\langle j^{(m)}| \tag{A25}$$

is the intra-graph adjacency matrix, i.e., within the complete graph $K_M^{(m)}$, and

$$A_{\text{inter}} = \sum_{m=1}^{M+1} \sum_{i=1}^{M} |i^{(m)}\rangle\langle (M+1-i)^{(m')}|, \tag{A26}$$

with $m' = 1 + \text{mod}(i + m - 1, M + 1)$, is the inter-graphs adjacency matrix, i.e., between different complete graphs. The index m labels the complete graphs $K_M^{(m)}$ forming the M-simplex. Note that Equation (A26) follows the labeling of the vertices in Figure A1 and it is just one of the possible ways to computationally implement the inter-graphs contribution.

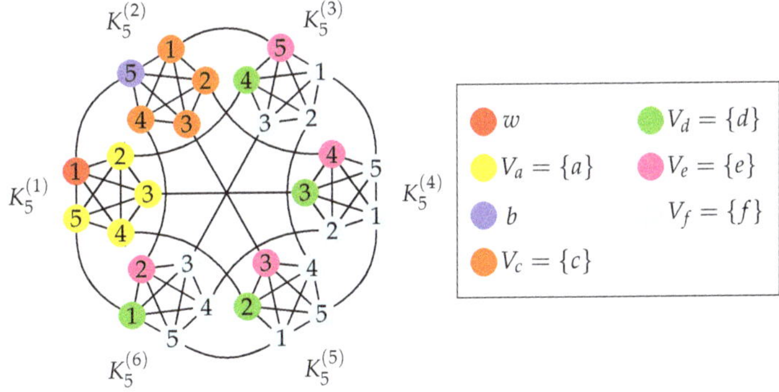

Figure A1. Labeling of vertices in a 5-simplex of complete graphs. The trap vertex w is colored red and assumed to be $|1\rangle$ in $K_5^{(1)}$. Same coloring denotes the subsets V_α of identically evolving vertices α, with $\alpha = w, a, b, c, d, e, f$ (see also Figure 6). Note that each of the two vertices w and b forms a subset of one element, itself.

In this case, using the notion of adjacency and reasoning by symmetry to introduce the subsets of the identically evolving vertices provide a framework which, analytically, is simpler and clearer to deal with than explicitly using the Laplacian above defined. These subsets contain the vertices which behave identically under the action of the Hamiltonian:

$$H|w\rangle = (M - i\kappa)|w\rangle - \sum_{i \in V_a}|i\rangle - |b\rangle, \tag{A27}$$

$$H\sum_{i \in V_a}|i\rangle = 2\sum_{i \in V_a}|i\rangle - (M-1)|w\rangle - \sum_{i \in V_d}|i\rangle, \tag{A28}$$

$$H|b\rangle = M|b\rangle - |w\rangle - \sum_{i \in V_c}|i\rangle, \tag{A29}$$

$$H\sum_{i \in V_c}|i\rangle = 2\sum_{i \in V_c}|i\rangle - (M-1)|b\rangle - \sum_{i \in V_e}|i\rangle, \tag{A30}$$

$$H\sum_{i \in V_d}|i\rangle = M\sum_{i \in V_d}|i\rangle - \sum_{i \in V_a}|i\rangle - \sum_{i \in V_e}|i\rangle - \sum_{i \in V_f}|i\rangle, \tag{A31}$$

$$H\sum_{i \in V_e}|i\rangle = M\sum_{i \in V_e}|i\rangle - \sum_{i \in V_c}|i\rangle - \sum_{i \in V_d}|i\rangle - \sum_{i \in V_f}|i\rangle, \tag{A32}$$

$$H\sum_{i \in V_f}|i\rangle = 2\sum_{i \in V_f}|i\rangle - (M-2)\left(\sum_{i \in V_d}|i\rangle + \sum_{i \in V_e}|i\rangle\right). \tag{A33}$$

Note that the results of H applied on the vertices in V_c or in V_d are different, and this is the reason why they form different subsets. According to these preliminary results, the basis states (33) are obtained, as follows:

$$H|e_1\rangle = (M - i\kappa)|w\rangle - \sum_{i \in V_a}|i\rangle - |b\rangle \xrightarrow{\text{O.N.}} |e_2\rangle, \tag{A34}$$

$$H|e_2\rangle = \frac{1}{\sqrt{M}}\left(2\sum_{i \in V_a}|i\rangle - M|w\rangle + M|b\rangle - \sum_{i \in V_c \cup V_d}|i\rangle\right) \xrightarrow{\text{O.N.}} |e_3\rangle, \tag{A35}$$

$$H|e_3\rangle = \frac{\sqrt{M}}{\sqrt{(M-1)(M^2 - 2M + 4)}}\left[\frac{M-4}{M}\sum_{i \in V_a}|i\rangle - (M-1)^2|b\rangle\right.$$

$$\left. + \frac{M^2 - M + 2}{M}\sum_{i \in V_c \cup V_d}|i\rangle - 2\sum_{i \in V_e}|i\rangle - \sum_{i \in V_f}|i\rangle\right] \xrightarrow{\text{O.N.}} |e_4\rangle, \tag{A36}$$

$$H|e_4\rangle = \frac{1}{\sqrt{(M-1)(M^2 - 2M + 4)(M^3 + 2M^2 - 8M + 16)}}\left\{(M^2 - 4)\left[\sum_{i \in V_a}|i\rangle - (M-1)|b\rangle\right]\right.$$

$$\left. + 2(M^2 - M + 2)\sum_{i \in V_c \cup V_d}|i\rangle - M(M^2 - 2M + 8)\sum_{i \in V_e}|i\rangle + (M-2)^2\sum_{i \in V_f}|i\rangle\right\} \xrightarrow{\text{O.N.}} |e_5\rangle, \tag{A37}$$

and it can be proved that

$$H|e_5\rangle = \frac{M+2}{M^3 + 2M^2 - 8M + 16}\left[M\sqrt{(M-2)(M^2 - 2M + 4)}|e_4\rangle + (M^3 - 4M + 8)|e_5\rangle\right]. \tag{A38}$$

In conclusion, any state $H^k|w\rangle \in \text{span}(\{|e_1\rangle, \dots, |e_5\rangle\}) \forall k \in \mathbb{N}_0$, thus the states (33) form an orthonormal basis for the subspace $\mathcal{I}(H, |w\rangle)$.

References

1. Farhi, E.; Gutmann, S. Quantum computation and decision trees. *Phys. Rev. A* **1998**, *58*, 915. [CrossRef]
2. Childs, A.M.; Farhi, E.; Gutmann, S. An example of the difference between quantum and classical random walks. *Quantum Inf. Process.* **2002**, *1*, 35–43. [CrossRef]
3. Wang, J.; Manouchehri, K. *Physical Implementation of Quantum Walks*; Springer: New York, NY, USA, 2013.
4. Du, J.; Li, H.; Xu, X.; Shi, M.; Wu, J.; Zhou, X.; Han, R. Experimental implementation of the quantum random-walk algorithm. *Phys. Rev. A* **2003**, *67*, 042316. [CrossRef]

5. Côté, R.; Russell, A.; Eyler, E.E.; Gould, P.L. Quantum random walk with Rydberg atoms in an optical lattice. *New J. Phys.* **2006**, *8*, 156. [CrossRef]
6. Qiang, X.; Loke, T.; Montanaro, A.; Aungskunsiri, K.; Zhou, X.; O'Brien, J.L.; Wang, J.B.; Matthews, J.C. Efficient quantum walk on a quantum processor. *Nat. Commun.* **2016**, *7*, 1–6.
7. Tang, H.; Lin, X.F.; Feng, Z.; Chen, J.Y.; Gao, J.; Sun, K.; Wang, C.Y.; Lai, P.C.; Xu, X.Y.; Wang, Y.; et al. Experimental two-dimensional quantum walk on a photonic chip. *Sci. Adv.* **2018**, *4*, eaat3174. [CrossRef]
8. Venegas-Andraca, S.E. *Quantum Walks for Computer Scientists*; Synthesis Lectures on Quantum Computing; Morgan & Claypool Publishers: San Rafael, CA, USA, 2008.
9. Portugal, R. *Quantum Walks and Search Algorithms*; Springer: Berlin/Heidelberg, Germany, 2018.
10. Childs, A.M.; Goldstone, J. Spatial search by quantum walk. *Phys. Rev. A* **2004**, *70*, 022314. [CrossRef]
11. Krok, M.; Rycerz, K.; Bubak, M. Application of Continuous Time Quantum Walks to Image Segmentation. In *International Conference on Computational Science*; Springer: Cham, Switzerland, 2019, pp. 17–30.
12. Lahini, Y.; Steinbrecher, G.R.; Bookatz, A.D.; Englund, D. Quantum logic using correlated one-dimensional quantum walks. *Npj Quantum Inf.* **2018**, *4*, 2. [CrossRef]
13. Childs, A.M. Universal computation by quantum walk. *Phys. Rev. Lett.* **2009**, *102*, 180501. [CrossRef]
14. Christandl, M.; Datta, N.; Ekert, A.; Landahl, A.J. Perfect state transfer in quantum spin networks. *Phys. Rev. Lett.* **2004**, *92*, 187902. [CrossRef]
15. Kendon, V.M.; Tamon, C. Perfect state transfer in quantum walks on graphs. *J. Comput. Theor. Nanosci.* **2011**, *8*, 422–433. [CrossRef]
16. Alvir, R.; Dever, S.; Lovitz, B.; Myer, J.; Tamon, C.; Xu, Y.; Zhan, H. Perfect state transfer in Laplacian quantum walk. *J. Algebr. Comb.* **2016**, *43*, 801–826. [CrossRef]
17. Lahini, Y.; Verbin, M.; Huber, S.D.; Bromberg, Y.; Pugatch, R.; Silberberg, Y. Quantum walk of two interacting bosons. *Phys. Rev. A* **2012**, *86*, 011603. [CrossRef]
18. Beggi, A.; Razzoli, L.; Bordone, P.; Paris, M.G.A. Probing the sign of the Hubbard interaction by two-particle quantum walks. *Phys. Rev. A* **2018**, *97*, 013610. [CrossRef]
19. Agliari, E.; Blumen, A.; Mülken, O. Dynamics of continuous-time quantum walks in restricted geometries. *J. Phys. A* **2008**, *41*, 445301. [CrossRef]
20. Salimi, S. Continuous-time quantum walks on semi-regular spidernet graphs via quantum probability theory. *Quantum Inf. Process.* **2010**, *9*, 75–91. [CrossRef]
21. Darázs, Z.; Anishchenko, A.; Kiss, T.; Blumen, A.; Mülken, O. Transport properties of continuous-time quantum walks on Sierpinski fractals. *Phys. Rev. E* **2014**, *90*, 032113. [CrossRef]
22. Li, X.; Chen, H.; Wu, M.; Ruan, Y.; Liu, Z.; Tan, J. Quantum transport on large-scale sparse regular networks by using continuous-time quantum walk. *Quantum Inf. Process.* **2020**, *19*, 1–13. [CrossRef]
23. Rai, A.; Agarwal, G.S.; Perk, J.H. Transport and quantum walk of nonclassical light in coupled waveguides. *Phys. Rev. A* **2008**, *78*, 042304. [CrossRef]
24. Blumen, A.; Bierbaum, V.; Mülken, O. Coherent dynamics on hierarchical systems. *Phys. A* **2006**, *371*, 10–15. [CrossRef]
25. Mülken, O.; Pernice, V.; Blumen, A. Quantum transport on small-world networks: A continuous-time quantum walk approach. *Phys. Rev. E* **2007**, *76*, 051125. [CrossRef] [PubMed]
26. Xu, X.P.; Li, W.; Liu, F. Coherent transport on Apollonian networks and continuous-time quantum walks. *Phys. Rev. E* **2008**, *78*, 052103. [CrossRef] [PubMed]
27. Yalouz, S.; Pouthier, V. Continuous-time quantum walk on an extended star graph: Trapping and superradiance transition. *Phys. Rev. E* **2018**, *97*, 022304. [CrossRef] [PubMed]
28. Mülken, O.; Blumen, A. Continuous-time quantum walks: Models for coherent transport on complex networks. *Phys. Rep.* **2011**, *502*, 37–87. [CrossRef]
29. Mülken, O.; Blumen, A.; Amthor, T.; Giese, C.; Reetz-Lamour, M.; Weidemüller, M. Survival probabilities in coherent exciton transfer with trapping. *Phys. Rev. Lett.* **2007**, *99*, 090601. [CrossRef] [PubMed]
30. Agliari, E.; Muelken, O.; Blumen, A. Continuous-time quantum walks and trapping. *Int. J. Bifurc. Chaos* **2010**, *20*, 271–279. [CrossRef]
31. Mülken, O.; Blumen, A. Slow transport by continuous time quantum walks. *Phys. Rev. E* **2005**, *71*, 016101. [CrossRef]
32. Mülken, O.; Blumen, A. Efficiency of quantum and classical transport on graphs. *Phys. Rev. E* **2006**, *73*, 066117. [CrossRef]
33. Lambert, N.; Chen, Y.N.; Cheng, Y.C.; Li, C.M.; Chen, G.Y.; Nori, F. Quantum biology. *Nat. Phys.* **2013**, *9*, 10–18. [CrossRef]
34. Mohseni, M.; Omar, Y.; Engel, G.S.; Plenio, M.B. *Quantum Effects in Biology*; Cambridge University Press: Cambridge, UK, 2014.
35. Mülken, O.; Bierbaum, V.; Blumen, A. Coherent exciton transport in dendrimers and continuous-time quantum walks. *J. Chem. Phys.* **2006**, *124*, 124905. [CrossRef]
36. Mohseni, M.; Rebentrost, P.; Lloyd, S.; Aspuru-Guzik, A. Environment-assisted quantum walks in photosynthetic energy transfer. *J. Chem. Phys.* **2008**, *129*, 174106. [CrossRef] [PubMed]
37. Rebentrost, P.; Mohseni, M.; Kassal, I.; Lloyd, S.; Aspuru-Guzik, A. Environment-assisted quantum transport. *New J. Phys.* **2009**, *11*, 033003. [CrossRef]

38. Plenio, M.B.; Huelga, S.F. Dephasing-assisted transport: Quantum networks and biomolecules. *New J. Phys.* **2008**, *10*, 113019. [CrossRef]

39. Olaya-Castro, A.; Lee, C.F.; Olsen, F.F.; Johnson, N.F. Efficiency of energy transfer in a light-harvesting system under quantum coherence. *Phys. Rev. B* **2008**, *78*, 085115. [CrossRef]

40. Caruso, F.; Chin, A.W.; Datta, A.; Huelga, S.F.; Plenio, M.B. Highly efficient energy excitation transfer in light-harvesting complexes: The fundamental role of noise-assisted transport. *J. Chem. Phys.* **2009**, *131*, 09B612. [CrossRef]

41. Hoyer, S.; Sarovar, M.; Whaley, K.B. Limits of quantum speedup in photosynthetic light harvesting. *New J. Phys.* **2010**, *12*, 065041. [CrossRef]

42. Novo, L.; Mohseni, M.; Omar, Y. Disorder-assisted quantum transport in suboptimal decoherence regimes. *Sci. Rep.* **2016**, *6*, 18142. [CrossRef]

43. Adronov, A.; Fréchet, J.M. Light-harvesting dendrimers. *Chem. Commun.* **2000**, *18*, 1701–1710.

44. Bradshaw, D.S.; Andrews, D.L. Mechanisms of light energy harvesting in dendrimers and hyperbranched polymers. *Polymers* **2011**, *3*, 2053–2077. [CrossRef]

45. Wong, T.G.; Tarrataca, L.; Nahimov, N. Laplacian versus adjacency matrix in quantum walk search. *Quantum Inf. Process.* **2016**, *15*, 4029–4048. [CrossRef]

46. Janmark, J.; Meyer, D.A.; Wong, T.G. Global symmetry is unnecessary for fast quantum search. *Phys. Rev. Lett.* **2014**, *112*, 210502. [CrossRef]

47. Meyer, D.A.; Wong, T.G. Connectivity is a poor indicator of fast quantum search. *Phys. Rev. Lett.* **2015**, *114*, 110503. [CrossRef] [PubMed]

48. Novo, L.; Chakraborty, S.; Mohseni, M.; Neven, H.; Omar, Y. Systematic dimensionality reduction for quantum walks: Optimal spatial search and transport on non-regular graphs. *Sci. Rep.* **2015**, *5*, 13304. [CrossRef] [PubMed]

49. Jafarizadeh, M.; Sufiani, R.; Salimi, S.; Jafarizadeh, S. Investigation of continuous-time quantum walk by using Krylov subspace-Lanczos algorithm. *Eur. Phys. J. B* **2007**, *59*, 199–216. [CrossRef]

50. Wong, T.G. Diagrammatic approach to quantum search. *Quantum Inf. Process.* **2015**, *14*, 1767–1775. [CrossRef]

51. Cameron, P.J.; Van Lint, J.H.; Cameron, P.J. *Designs, Graphs, Codes and Their Links*; Cambridge University Press: Cambridge, UK, 1991; Volume 3.

52. Brouwer, A.E.; Haemers, W.H. *Spectra of Graphs*; Springer Science & Business Media: Berlin/Heidelberg, Germany, 2011.

53. Weisstein, E.W. Prime Power. From MathWorld—A Wolfram Web Resource. Available online: https://mathworld.wolfram.com/PrimePower.html (accessed on 20 October 2020).

54. Nelson, D.R.; Fisher, M.E. Soluble renormalization groups and scaling fields for low-dimensional Ising systems. *Ann. Phys.* **1975**, *91*, 226–274. [CrossRef]

55. Dhar, D. Lattices of effectively nonintegral dimensionality. *J. Math. Phys.* **1977**, *18*, 577–585. [CrossRef]

56. Dhar, D. Self-avoiding random walks: Some exactly soluble cases. *J. Math. Phys.* **1978**, *19*, 5–11. [CrossRef]

57. Wang, Y.; Wu, S.; Wang, W. Optimal quantum search on truncated simplex lattices. *Phys. Rev. A* **2020**, *101*, 062333. [CrossRef]

58. Fiedler, M. Laplacian of graphs and algebraic connectivity. *Banach Cent. Publ.* **1989**, *25*, 57–70. [CrossRef]

59. West, D.B. *Introduction to Graph Theory*, 2nd ed.; Prentice Hall: Upper Saddle River, NJ, USA, 2001.

60. Fiedler, M. Algebraic connectivity of graphs. *Czechoslov. Math. J.* **1973**, *23*, 298–305. [CrossRef]

61. De Abreu, N.M.M. Old and new results on algebraic connectivity of graphs. *Linear Algebra Its Appl.* **2007**, *423*, 53–73. [CrossRef]

62. Beineke, L.W.; Wilson, R.J.; Cameron, P.J. (Eds.) *Topics in Algebraic Graph Theory*; Cambridge University Press: Cambridge, UK, 2004; Volume 102.

63. Chartrand, G.; Zhang, P. *A First Course in Graph Theory*; Dover Publications: Mineola, NY, USA, 2012.

64. Chung, F.R.; Graham, F.C. *Spectral Graph Theory*; Number 92; American Mathematical Soc.: Providence, RI, USA, 1997.

65. Banerjee, S.; Roy, A. *Linear Algebra and Matrix Analysis for Statistics*; CRC Press: Boca Raton, FL, USA, 2014.

Article

Scattering as a Quantum Metrology Problem: A Quantum Walk Approach

Francesco Zatelli [1], Claudia Benedetti [1,*] and Matteo G. A. Paris [1,2]

[1] Dipartimento di Fisica 'Aldo Pontremoli', Università degli Studi di Milano, I-20133 Milano, Italy; francesco.zatelli@gmail.com (F.Z.); matteo.paris@fisica.unimi.it (M.G.A.P.)

[2] INFN, Sezione di Milano, I-20133 Milano, Italy

* Correspondence: claudia.benedetti@unimi.it

Received: 14 October 2020; Accepted: 17 November 2020; Published: 19 November 2020

Abstract: We address the scattering of a quantum particle by a one-dimensional barrier potential over a set of discrete positions. We formalize the problem as a continuous-time quantum walk on a lattice with an impurity and use the quantum Fisher information as a means to quantify the maximal possible accuracy in the estimation of the height of the barrier. We introduce suitable initial states of the walker and derive the reflection and transmission probabilities of the scattered state. We show that while the quantum Fisher information is affected by the width and central momentum of the initial wave packet, this dependency is weaker for the quantum signal-to-noise ratio. We also show that a dichotomic position measurement provides a nearly optimal detection scheme.

Keywords: quantum walks; scattering; quantum metrology; quantum Fisher information; optimal measurement

---

1. Introduction

Since the Rutherford experiment [1], scattering has played a central role in the study of unknown interactions in many fields of physics [2–4]. At its core, a scattering experiment may be viewed as a parameter-estimation problem. Indeed, the scattering potential can be modeled with a set of unknown parameters that characterize the evolution of the quantum particles that impinge on it. Estimating the value of those parameters then involves measurements that are performed on the scattered state, followed by a collection of outputs that are used to build estimators for the parameters. If we consider scattering as an estimation problem, we can study the maximum amount of information that can be extracted from a single measurement on the quantum system, and we can assess the performance of feasible detection schemes. All these questions find answers in the theory of local quantum estimation, which has the aim of quantifying the best precision of an estimation procedure [5]. Indeed, in the past few years, local quantum estimation theory has been applied to a variety of problems, such as the estimation of the relevant parameters of quantum structured baths [6–10], graph and lattice properties [11–13], and classical processes [14].

In this work, we analyze the one-dimensional scattering of a quantum particle from a potential barrier with the aim of inferring its height. The particle moves on a set of discrete positions, and it is thus described as a continuous-time quantum walk (CT QW) on the line with a central barrier. The barrier is implemented by a detuning of the energy of the central site with respect to the other sites. As a matter of fact, the analysis of the evolution of a quantum walk in the presence of a barrier is strongly connected with the study of defects and impurities in implementations of QW [15–18]. A detuning in the on-site energy of a site can be interpreted as a defect that influences the dynamics and the scattering properties of the walker. Understanding the role of imperfections is of fundamental importance for a realistic description of the QWs. In fact, knowing how a protocol or an algorithm [19–23] is affected by impurities and noise allows us to hinder or even neutralize detrimental effects.

Inspired by previous works on the discretization of continuous-systems [24,25], we first derive scattered states on the infinite line of discrete positions. In order to consider physically relevant states for the walker, we initialize the particle in a Gaussian wave packet with central initial momentum k_0 and standard deviation σ. We evaluate the transmission probability through the barrier and the maximum extractable information as a function of these two free parameters. We show that the quantum Fisher information (QFI) is strongly affected by the value of the initial central momentum of the walker, but only slightly by the initial spread of the wave packet. The quantum signal-to-noise ratio has a maximum corresponding to the optimal value of the barrier height that can be better estimated. Finally, we consider a feasible measurement, i.e., a dichotomic position measurement, and we compare its Fisher information (FI) with the QFI. We show that this measurement is nearly optimal, i.e., its FI is close to the QFI in almost all the parameter space we consider.

The paper is organized as follows: In Section 2, we introduce the concept of CTQW with inhomogeneous on-site energies, and in Section 3, we briefly review the main concepts of local quantum estimation theory. In Section 4, we introduce the free-particle scattering states, and then, we use them to build the physically relevant wave packets, whose transmission and reflection probabilities are derived. In Section 5, we compute the QFI for initial Gaussian wave packets, and we compare its value with the FI of a dichotomic position measurement. Finally, in Section 6, we draw our conclusions.

2. Quantum Walks with Inhomogeneous On-Site Energies

A CTQW model describes the evolution of a quantum particle over a discrete set of positions, continuously in time [26,27]. It evolves in an N-dimensional Hilbert space with orthonormal basis states $\{|j\rangle\}_{j\in\mathbb{Z}}$, which represent the positions that can be occupied by the walker. The Hamiltonian of a CTQW on the line with inhomogeneous on-site energies ϵ_j and uniform couplings J_0 has the expression ($\hbar = 1$):

$$H = \sum_j \epsilon_j \,|j\rangle\langle j| - J_0 \sum_j \Big(|j\rangle\langle j+1| + |j+1\rangle\langle j| \Big). \tag{1}$$

Without loss of generality, we fix $J_0 = 1$, thus expressing time and ϵ_j in units of J_0. If we set $\epsilon_j = 2\,\forall j$, we recover the graph Laplacian L, i.e., $H = -L$. It is worth mentioning that for the one-dimensional lattice, L represents the discretized version of Laplace operator ∇^2, and $-L$ is kinetic energy operator of a particle with mass $m = \frac{1}{2}$ constrained to a discrete set of positions [28].

Given a set of on-site energies $\{\epsilon_j\}$, it is possible to separate the Hamiltonian into a kinetic and a potential operator, L and V respectively. The Hamiltonian can thus be written as $H = -L + V$ with:

$$L = \sum_j \Big[-2\,|j\rangle\langle j| + |j\rangle\langle j+1| + |j+1\rangle\langle j| \Big] \quad \text{and} \quad V = \sum_j V_j \,|j\rangle\langle j| = \sum_j (\epsilon_j - 2)\,|j\rangle\langle j| \tag{2}$$

highlighting the fact that for $\epsilon_j = 2\,\forall j$, the unperturbed Laplacian Hamiltonian is obtained. Due to the tridiagonal form of the matrix H, the eigenvalue equation $H|\psi^{(k)}\rangle = E_k|\psi^{(k)}\rangle$ can be recast in the form of a three-term recurrence relation. By explicitly writing H in terms of the Laplacian and potential parts and projecting into a basis state $|j\rangle$, we obtain $\langle j| -L + V|\psi^{(k)}\rangle = E_k\,\langle j|\,\psi^{(k)}\rangle$ and the recurrence relation:

$$-\psi_{j+1}^{(k)} + 2\psi_j^{(k)} - \psi_{j-1}^{(k)} + V_j\psi_j^{(k)} = E_k\psi_j^{(k)}, \tag{3}$$

where $|\psi^{(k)}\rangle = \sum_j \psi_j^{(k)} |j\rangle$. Equation (3) is easily identifiable with the discretization in the position basis of the time-independent Schrödinger equation for a particle of mass $m = \frac{1}{2}$.

In analogy with the continuous case, we introduce the momentum states as the Fourier series of the countable orthonormal set of position eigenstates. In particular, we define the momentum state $|k\rangle$ through a discrete-time Fourier transform (DTFT):

$$|k\rangle = \frac{1}{\sqrt{2\pi}} \sum_{j \in \mathbb{Z}} e^{ikj} |j\rangle, \quad k \in (-\pi, \pi] \tag{4}$$

$$|j\rangle = \frac{1}{\sqrt{2\pi}} \int_{-\pi}^{\pi} e^{-ikj} |k\rangle \, dk, \quad j \in \mathbb{Z}. \tag{5}$$

If no external potential is considered, i.e., $V_j = 0 \, \forall j$, the states $\{|k\rangle\}$ are solutions to Equation (3) with $\psi_j^{(k)} = e^{ikj}$ and corresponding energies $E_k = 2 - 2\cos(k)$. The dispersion relation implies that the phase velocity v_p and the group velocity v_g are:

$$v_p = \frac{E_k}{k} = \frac{2 - 2\cos(k)}{k}, \quad v_g = \frac{\partial E_k}{\partial k} = 2\sin(k). \tag{6}$$

Thus, the momentum states (4) are the discretization of the plane waves with the dispersion relation typical of the tight-binding models [29]. We identify these states as free particle states because, in analogy with the continuous case, plane waves are the eigenstates of a purely kinetic Hamiltonian. This suggests that the separation of the QW Hamiltonian into a kinetic term and a potential one is indeed meaningful. In the following, we are going to introduce an obstacle, i.e., an external potential that causes an inhomogeneity on the on-site energies.

3. Tools of Local Quantum Estimation Theory

Before analyzing the QW scattering from a barrier, we review a few key concepts in the theory of local quantum estimation. Consider a sample of M independent outcomes of a measurement $\{x_1, x_2, \ldots, x_M\}$ drawn from the probability distribution $p(x|\Delta)$, where Δ is an unknown parameter we wish to estimate. The Cramèr–Rao (CR) inequality imposes a lower bound on the variance of any unbiased estimator $\hat{\Delta}(\{x_1, x_2, \ldots, x_M\})$ for such a parameter:

$$\mathrm{Var}(\hat{\Delta}) \geq \frac{1}{MF(\Delta)} \tag{7}$$

where $F(\Delta)$ is the Fisher information, defined as:

$$F(\Delta) = \int \left(\frac{\partial \ln p(x|\Delta)}{\partial \Delta} \right)^2 p(x|\Delta) \, dx = \int \left(\frac{\partial p(x|\Delta)}{\partial \Delta} \right)^2 \frac{1}{p(x|\Delta)} \, dx. \tag{8}$$

The quantum version of the CR bound is derived by generalizing the concept of FI. This is done by maximizing the FI over all possible measurements, and the obtained quantity is called quantum Fisher information $H(\Delta)$. A detailed derivation of the QFI can be found in [30]. The quantum CR bound takes the following form:

$$\mathrm{Var}(\hat{\Delta}) \geq \frac{1}{MH(\Delta)}. \tag{9}$$

and follows from the inequality $F(\Delta) \leq H(\Delta)$, which provides the basis for the identification of the QFI with the ultimate bound to precision of any unbiased estimator. The aim of local quantum estimation theory is to determine the maximum extractable information from a quantum probe, whose state depends on the value of the parameter. If only pure states are considered as probes, i.e., a parameter-dependent family of quantum states $|\psi_\Delta\rangle$, the QFI can be explicitly written as [30]:

$$H(\Delta) = 4\left[\langle\partial_\Delta\psi_\Delta|\partial_\Delta\psi_\Delta\rangle - |\langle\psi_\Delta|\partial_\Delta\psi_\Delta\rangle|^2\right],\tag{10}$$

where $|\partial_\Delta\psi_\Delta\rangle$ represents the derivative of the state with respect to the parameter Δ. A suitable figure of merit that can be used in order to evaluate the estimability of a parameter is the quantum signal-to-noise ratio (QSNR):

$$R(\Delta) = \Delta^2 H(\Delta),\tag{11}$$

which provides an upper bound to the signal-to-noise ratio $\hat{\Delta}^2/\operatorname{Var}(\hat{\Delta})$ of any detection scheme.

4. Scattering in the Presence of an Obstacle

Let us now consider a situation where there is an obstacle placed in the middle of the chain. The obstacle, or barrier, has the width of a single site, i.e., all sites have the same energy $\epsilon_j = 2$, except for the central one $|0\rangle$, which has a detuning Δ, such that $\epsilon_0 = 2 + \Delta$. Thus, the Hamiltonian defined in Equation (1) is modified by placing the obstacle at $j = 0$, and it becomes:

$$H = \sum_{j\in\mathbb{Z}}\left(2\,|j\rangle\langle j| - |j+1\rangle\langle j| - |j\rangle\langle j+1|\right) + \Delta\,|0\rangle\langle 0|.\tag{12}$$

The site $j = 0$ has on-site energy $\epsilon_0 = 2 + \Delta$ or, alternatively said, potential $V_0 = \Delta$. In order to study the scattering properties of such model, we start by deriving the scattering states.

4.1. Scattering States

Scattering states for one-dimensional systems in the continuous-space case are known for a variety of potentials [31]. We now want to derive such states for the discrete system under consideration. The generic stationary scattering state $|\psi_s\rangle$ with fixed momentum k can be written as a linear combination of free particle states, namely:

$$\langle j|\psi_s\rangle = \begin{cases} Ae^{ikj} + Be^{-ikj}, & j \leq 0 \\ Ce^{ikj}, & j \geq 0 \end{cases},\tag{13}$$

where the terms proportional to A, B, and C correspond to the incident, the reflected, and the transmitted wave, respectively. The coefficients are calculated imposing that the two parts of the state (before and after the obstacle) are properly connected at $j = 0$, i.e., by discretizing the continuity conditions, and using the recurrence relations (3), i.e., $\langle-1|\psi_s\rangle - \Delta\langle 0|\psi_s\rangle + \langle 1|\psi_s\rangle = 2\cos(k)\langle 0|\psi_s\rangle$, which represent the discontinuity introduced by the obstacle. Therefore, the reflection $R = \frac{|B|^2}{|A|^2}$ and transmission $T = \frac{|C|^2}{|A|^2}$ coefficients can be easily calculated through:

$$\begin{cases} A + B = C \\ Ae^{-ik} + Be^{ik} = C(2\cos(k) + \Delta - e^{ik}) \end{cases} \longrightarrow \begin{cases} B = \dfrac{1}{\frac{2i\sin k}{\Delta} - 1}A \\ C = \dfrac{1}{1 - \frac{\Delta}{2i\sin k}}A \end{cases},\tag{14}$$

and they have the expressions:

$$R(\Delta, k) = \frac{1}{1 + \dfrac{4\sin^2(k)}{\Delta^2}},\quad T(\Delta, k) = \frac{1}{1 + \dfrac{\Delta^2}{4\sin^2(k)}}.\tag{15}$$

These coefficients closely resemble those corresponding to a delta potential in a continuous system [31]; in particular, the coefficients only depend on Δ^2, meaning that there is no difference between an attractive or repulsive potential as concerns scattering. If Δ is fixed, T is maximum for $k = \frac{\pi}{2}$, which corresponds to the highest group velocity (but not to the highest energy). Consistently, at the same value of k, R has a minimum. As the absolute value of Δ is increased, the transmission coefficient drops to smaller values, as reported in Figure 1. For every incident $|k\rangle$, we may thus define:

$$S|k\rangle = \frac{B}{A}|-k\rangle + \frac{C}{A}|k\rangle \tag{16}$$

where we introduced a scattering matrix S whose elements give information on the reflection and transmission coefficients [31]. If we set $A = |A|$ and we highlight the phases of the reflected and transmitted waves, we obtain:

$$S|k\rangle = \frac{|B|}{|A|}e^{i\phi_B}|-k\rangle + \frac{|C|}{|A|}e^{i\phi_C}|k\rangle = e^{i\phi_B}\left(\sqrt{R(\Delta,k)}\,|-k\rangle + \sqrt{T(\Delta,k)}e^{i(\phi_C-\phi_B)}\,|k\rangle\right). \tag{17}$$

The relative phase $e^{i(\phi_C-\phi_B)}$ can be computed from the ratio $\frac{C}{B}$ from Equation (14) and is equal to $\pi/2$. It follows that:

$$S|k\rangle = e^{i\phi_B(\Delta,k)}\left(\sqrt{R(\Delta,k)}\,|-k\rangle + i\sqrt{T(\Delta,k)}\,|k\rangle\right), \tag{18}$$

with the phase $\phi_B(\Delta,k) = \arctan\left(\frac{2\sin(k)}{\Delta}\right)$.

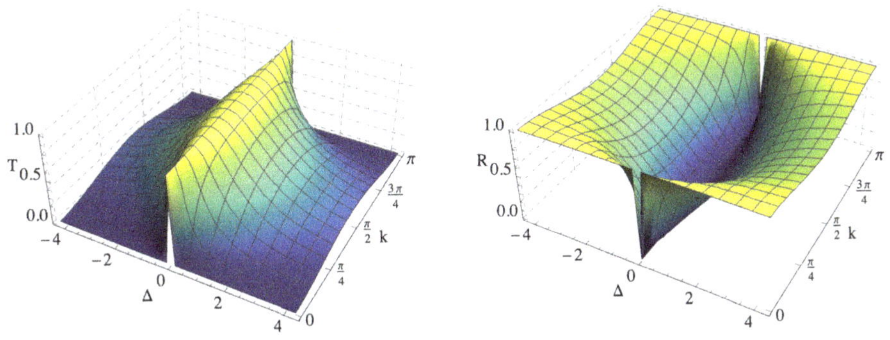

Figure 1. Transmission and reflection coefficients T and R as a function of Δ and k.

It is possible to define the reflection and transmission coefficients for more general states. Given an initial localized wave packed $|\psi_0\rangle$ placed on the left of the obstacle, its time-evolved state is:

$$|\psi(t)\rangle = e^{-iHt}|\psi_0\rangle. \tag{19}$$

We define the time-dependent probabilities:

$$\rho(t) = \sum_{j<0}|\langle j|\psi(t)\rangle|^2, \qquad \tau(t) = \sum_{j>0}|\langle j|\psi(t)\rangle|^2, \qquad \delta(t) = |\langle 0|\psi(t)\rangle|^2. \tag{20}$$

The quantities $\rho(t)$ and $\tau(t)$ are indeed the probability of finding the walker before and after the obstacle, respectively. The defect coefficient $\delta(t)$ is the remaining probability, namely the probability of finding the particle on the obstacle site. In particular, when the scattering is over, the coefficient $\delta(t)$ is expected to vanish, and consequently, $\rho(t) + \tau(t) = 1$.

4.2. Gaussian Wave Packets

The vector described by Equation (18) is the mathematical building block from which we derive the asymptotic values of the quantities of interest; however, it is not normalizable and does not represent a physical state. For this reason, we now introduce more realistic states that are spatially localized. In particular, we consider a discretized version of a Gaussian wave packet:

$$\left| \mathcal{G}_{k_0} \right\rangle = \mathcal{N} \sum_{j \in \mathbb{Z}} e^{-\frac{(j-\mu)^2}{2\sigma^2}} e^{ik_0 j} \left| j \right\rangle. \tag{21}$$

The probability distribution of this state is a discretized Gaussian function with mean μ and variance $\frac{\sigma^2}{2}$. $\mathcal{N}$ is a normalization constant, while the parameter $k_0 \in (-\pi, \pi]$ represents the mean of the probability distribution in the momentum basis. The $\left| \mathcal{G}_{k_0} \right\rangle$ state in the momentum basis is still Gaussian under proper assumptions, and it has the expression:

$$\left| \mathcal{G}_{k_0} \right\rangle = \int_{-\pi}^{\pi} g_{k_0}(k) \left| k \right\rangle dk, \tag{22}$$

$$\text{with} \quad g_{k_0}(k) = \left\langle k \middle| \mathcal{G}_{k_0} \right\rangle \approx \sqrt{\frac{\sigma}{\pi^{1/2}}} e^{-\frac{(k-k_0)^2 \sigma^2}{2}} e^{-i\mu k}. \tag{23}$$

The detailed derivation of Expression (23) is shown in Appendix A. The crucial approximation made to obtain this expression is to consider narrow wave packets in the reciprocal space (i.e., $\min(|k_0 + \pi|, |\pi - k_0|) \gg 1/\sigma$; see Appendix A). Therefore, the Fourier transform of the Gaussian wave packet is not exactly a Gaussian in the momentum basis. Nevertheless, if the transformed state is sufficiently localized in reciprocal space, Equation (23) is a reasonable approximation.

4.3. Scattering with Gaussian Wave Packets

Here, we want to analyze the asymptotic scattering properties of an incident Gaussian wave packet. In order to do so, we exploit the results obtained for single momentum states $\left| k \right\rangle$. The Gaussian state in the momentum basis has the expression (22) where the Gaussian weights are included in $g_{k_0}(k)$. We consider a wave packet incident on the obstacle from the left ($j < 0$). Using (18) and linearity, the scattered Gaussian state can be written in the asymptotic limit as:

$$
\begin{aligned}
\left| \psi_{k_0, \Delta} \right\rangle = S \left| \mathcal{G}_{k_0} \right\rangle &= \int_{-\pi}^{\pi} g_{k_0}(k) S \left| k \right\rangle dk \\
&= \int_{-\pi}^{\pi} g_{k_0}(k) e^{i\phi_B(\Delta, k)} \left(\sqrt{R(\Delta, k)} \left| -k \right\rangle + i\sqrt{T(\Delta, k)} \left| k \right\rangle \right) dk \\
&= \int_{-\pi}^{\pi} \left(e^{-i\phi_B(\Delta, k)} \sqrt{R(\Delta, k)} |g_{-k_0}(k)| e^{i\mu k} + e^{i\phi_B(\Delta, k)} i\sqrt{T(\Delta, k)} |g_{k_0}(k)| e^{-i\mu k} \right) \left| k \right\rangle dk,
\end{aligned} \tag{24}
$$

where, in the last line, we used the equalities $|g_{k_0}(-k)| = |g_{-k_0}(k)|$, $R(\Delta, k) = R(\Delta, -k)$, and $\phi_B(\Delta, k) = -\phi_B(\Delta, -k)$. By inspection of Equation (24), we learn that the original Gaussian wave packet is divided into the superposition of two wave packets centered around opposite values of momentum k_0 and $-k_0$, corresponding to the transmitted and reflected wave function, respectively. These two wave packets are not Gaussian anymore, since they are weighted with scattering coefficients that depend on k. It is important to highlight that this description fails if the two wave packets overlap, which can happen if the original state is spread in k-space or if its mean is $k_0 \approx 0$ (or any multiple of π). The assumption of a narrow initial wave packet in k-space was already imposed in order to derive Equation (23), while asking for a $k_0 \neq 0$ corresponds to considering a wave packet with the group velocity different from zero. With these assumptions, the transmission and reflection coefficients can be calculated considering the probabilities of the reflected and transmitted wave packets:

$$\rho_{\mathcal{G}}(k_0, \Delta) = \int_{-\pi}^{\pi} R(\Delta, k) \left| g_{k_0}(k) \right|^2 dk, \quad \tau_{\mathcal{G}}(k_0, \Delta) = \int_{-\pi}^{\pi} T(\Delta, k) \left| g_{k_0}(k) \right|^2 dk. \tag{25}$$

This results are confirmed by numerical evaluation of the $\rho(t)$ and $\tau(t)$ coefficients in Equation (20) and shown in Figure 2. The dynamics of the walker is computed through Equation (19) for fixed values of k_0 and Δ. The figure shows that at long times, i.e., in the asymptotic limit, the transmission probability achieves exactly $\tau_{\mathcal{G}}(k_0, \Delta)$. A large transmission probability is associated with high values of k_0 and small values of Δ, while a small initial central momentum and a large barrier prevent good transmission.

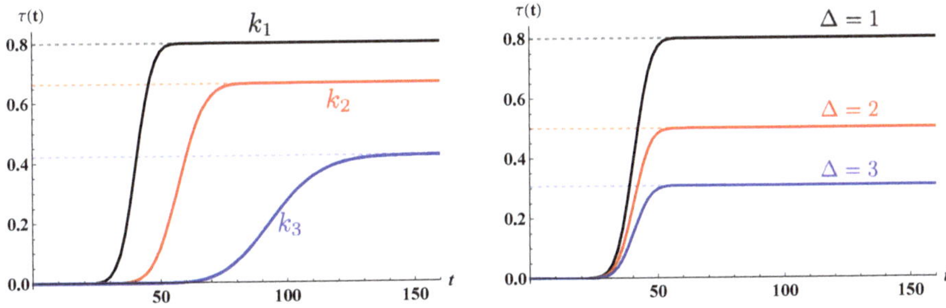

Figure 2. Transmission probability $\tau(t)$. The left plot is for a fixed value of $\Delta = 1$ and for decreasing values of $k_0 = k_1$, k_2, k_3, with $k_1 = \pi/2$ (black), $k_2 = \pi/4$ (red), $k_3 = \pi/7$ (blue). In the right plot, $k_0 = \pi/2$ is kept fixed while varying the disorder $\Delta = 1$ (black), $\Delta = 2$ (red), $\Delta = 3$ (blue). The dashed lines correspond to the value of the transmission coefficient $\tau_{\mathcal{G}}(k_0, \Delta)$ in Equation (25). In both plots, we considered $\sigma = 15$.

5. Quantum Estimation of a Scattering Potential

After having derived the scattered expression of a Gaussian wave packet, we turn our attention to the optimal estimation of the barrier height, i.e., of the parameter Δ. In order to do so, we prepare an initial Gaussian wave packet with initial central momentum k_0. In a scattering experiment, measurements can be performed only on the scattered state, which has the expression of Equation (24), which we report here for convenience:

$$\left| \psi_{k_0, \Delta} \right\rangle = \int_{-\pi}^{\pi} g_{k_0}(k) e^{i\phi_B(\Delta, k)} \left(\sqrt{R(\Delta, k)} \left| -k \right\rangle + i\sqrt{T(\Delta, k)} \left| k \right\rangle \right) dk.$$

In order to compute the QFI, Equation (10), we need the derivative:

$$\left| \partial_\Delta \psi_{k_0, \Delta} \right\rangle = \int_{-\pi}^{\pi} g_{k_0}(k) e^{i\phi_B(\Delta, k)} \times$$

$$\times \left[i\partial_\Delta \phi_B(\Delta, k) \left(\sqrt{R(\Delta, k)} \left| -k \right\rangle + i\sqrt{T(\Delta, k)} \left| k \right\rangle \right) + \left(\frac{\partial_\Delta R(\Delta, k)}{2\sqrt{R(\Delta, k)}} \left| -k \right\rangle + i\frac{\partial_\Delta T(\Delta, k)}{2\sqrt{T(\Delta, k)}} \left| k \right\rangle \right) \right] dk,$$

and the inner products:

$$\langle \partial_\Delta \psi_{k_0,\Delta} | \partial_\Delta \psi_{k_0,\Delta} \rangle = \int_{-\pi}^{\pi} |g_{k_0}(k)|^2 \left([\partial_\Delta \phi_B(\Delta,k)]^2 + \frac{[\partial_\Delta R(\Delta,k)]^2}{4\,R(\Delta,k)} + \frac{[\partial_\Delta T(\Delta,k)]^2}{4\,T(\Delta,k)} \right) dk \qquad (26)$$

$$\langle \psi_{k_0,\Delta} | \partial_\Delta \psi_{k_0,\Delta} \rangle = i \int_{-\pi}^{\pi} |g_{k_0}(k)|^2 \partial_\Delta \phi_B(\Delta,k)\, dk, \qquad (27)$$

with $\partial_\Delta R(\Delta,k) + \partial_\Delta T(\Delta,k) = 0$. We remind the reader that in this work, we always assume that the reflected and transmitted wave packets of the post-scattering state do not overlap, neither in position nor in momentum space. Notice that with this assumption, we also exclude slow states, i.e., those states with $k_0 \approx 0$ or $k_0 \approx \pi$. The QFI for an initial Gaussian wave packet may be computed through Equation (10):

$$H_{\mathcal{G}}(k_0,\Delta) = \int_{-\pi}^{\pi} |g_{k_0}(k)|^2 \left(\frac{[\partial_\Delta R(\Delta,k)]^2}{R(\Delta,k)} + \frac{[\partial_\Delta T(\Delta,k)]^2}{T(\Delta,k)} + 4[\partial_\Delta \phi_B(\Delta,k)]^2 \right) dk$$

$$- 4 \left(\int_{-\pi}^{\pi} |g_{k_0}(k)|^2 \partial_\Delta \phi_B(\Delta,k) dk \right)^2 \qquad (28)$$

$$= \frac{16 \sin^2 k_0}{[2 + \Delta^2 - 2\cos(2k_0)]^2} + \frac{g_H(k_0,\Delta)}{\sigma^2} + O(1/\sigma^3), \qquad (29)$$

where the explicit expression of $g_H(k_0,\Delta)$ is reported in Appendix B.

The typical behavior of the QFI as a function of Δ and the initial central momentum k_0 is shown in Figure 3. Since we want to avoid overlaps of the reflected and transmitted wave functions in momentum space, we exclude values for k_0 in the neighborhood of $k_0 = 0$ and $k_0 = \pi$. The QFI is symmetric under the exchange of the sign of the barrier, i.e., $\Delta \to -\Delta$, and it has a maximum centered in $\Delta = 0$. Small values of the barrier height $|\Delta| \ll 1$ have a larger QFI with respect to higher barriers. The spread of the wave packet σ affects the maximum precision only for $|\Delta| \ll 1$, as shown in the upper panel of Figure 4. From these plots, we can also see that the initial central momentum has an important role: in fact, as k_0 is increased from small values to $\frac{\pi}{2}$, the maximum of the QFI decreases.

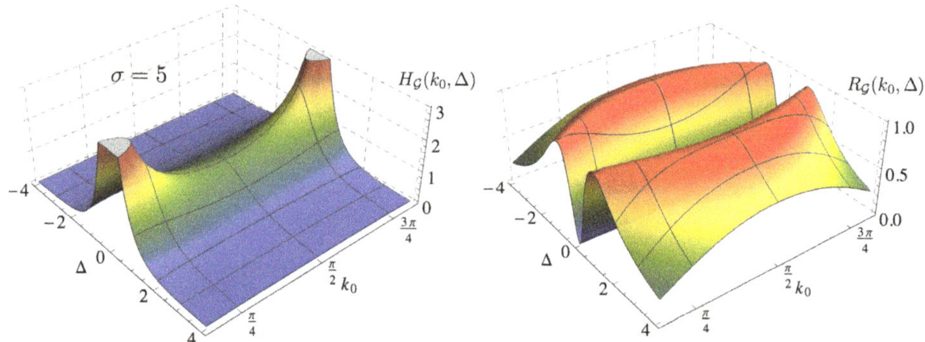

Figure 3. Left: QFI $H_{\mathcal{G}}(k_0,\Delta)$ for an initial Gaussian wave packet with $\sigma = 5$. Right: QSNR $R_{\mathcal{G}}(k_0,\Delta)$ for the same initial Gaussian wave packet.

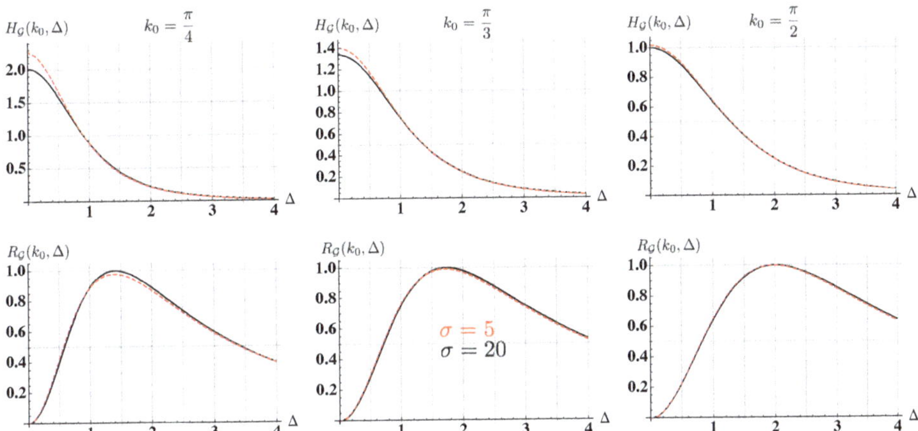

Figure 4. Comparison between the QFI (upper panel) and the QSNR (lower panel) with a large and a narrow initial wave packet in k-space, as a function of Δ and for three different values of k_0. The black solid lines are for $\sigma = 20$, while the dashed red lines are for $\sigma = 5$. The considered values of initial momentum are $k_0 = \frac{\pi}{4}, \frac{\pi}{3}, \frac{\pi}{2}$ for the left, center, and right column, respectively.

In order to compare the error of an estimator with the true value of the parameter to be estimated, we also address the QSNR, defined in Equation (11). Its behavior is shown in the right plot of Figure 3 and in the lower panel of Figure 4, for three different values of the initial central momentum k_0. The QSNR has a maximum for $\Delta \neq 0$, which corresponds to the value of the barrier height that can be better estimated. As the value of the initial central momentum is increased toward $k_0 = \frac{\pi}{2}$, the value of the optimal Δ slightly increases. The dependency on σ is negligible when considering the QSNR, as shown in the lower plots, where the behaviors for $\sigma = 5$ and $\sigma = 20$ are compared. Quite remarkably, the maximum value of the QSNR is very similar, $R_{\mathcal{G}} \approx 1$ for the considered values of k_0, thus making the initial central momentum a tool to fine tune the optimal value of Δ, but not the corresponding precision.

The behavior of the QSNR has an intuitive and straightforward physical interpretation. If the height of the barrier is negligible ($\Delta \ll 1$), then the walker is mostly transmitted anyway, and it is very difficult to detect small variations of Δ itself. Similarly, if $\Delta \gg 1$, the walker is mostly reflected independently of the exact value of Δ. On the other hand, for intermediate values of Δ, the wavefunction of the walker is very sensitive to its value, and measuring the walker indeed provides information. This picture is confirmed if one looks at the zeroth order expression of the QFI in Equation (29), which says that the maxima of the QSNR are located at $\Delta^2 = 2[1 - \cos(2k_0)]$. Notice that the values of (Δ, k_0) satisfying this relations are those making the reflection and transmission equal to each other $R(\sqrt{2[1 - \cos(2k_0)]}, k_0) = T(\sqrt{2[1 - \cos(2k_0)]}, k_0) = \frac{1}{2}$.

Dichotomic Position Measurement

We now address the question of whether a realistic position measurement is optimal, i.e., its FI equals the QFI defined in Equation (28). In particular, we consider a dichotomic measurement that just tells us if the particle is located on the left or on the right side of the barrier. Since we know from the Equation (25) that the quantities $\rho_{\mathcal{G}}(k_0, \Delta)$ and $\tau_{\mathcal{G}}(k_0, \Delta)$ correspond to the probabilities of finding the particle before or after the obstacle, the FI takes the expression:

$$F_{\mathcal{G}}(k_0, \Delta) = \frac{[\partial_\Delta \rho_{\mathcal{G}}(k_0, \Delta)]^2}{\rho_{\mathcal{G}}(k_0, \Delta)} + \frac{[\partial_\Delta \tau_{\mathcal{G}}(k_0, \Delta)]^2}{\tau_{\mathcal{G}}(k_0, \Delta)} = \frac{[\partial_\Delta \tau_{\mathcal{G}}(k_0, \Delta)]^2}{\tau_{\mathcal{G}}(k_0, \Delta)[1 - \tau_{\mathcal{G}}(k_0, \Delta)]} \tag{30}$$

$$= \frac{16 \sin^2 k_0}{[2 + \Delta^2 - 2\cos(2k_0)]^2} + \frac{g_F(k_0, \Delta)}{\sigma^2} + O(1/\sigma^3), \tag{31}$$

where the explicit expression of $g_F(k_0, \Delta)$ is reported in Appendix B. As the value of σ is increased, i.e., the wave packet is more localized in k-space, the FI of the dichotomic measurement approaches the QFI. The second order coefficients $g_s(k_0, \Delta)$, $s = H, F$ are different for the QFI and the FI (see Appendix B), but in the range of parameters we explore ($\sigma > 5$, $0 < \Delta \leq 4$, $0 < k_0 < \pi$), the ratio $\gamma(k_0, \Delta) = F_{\mathcal{G}}(k_0, \Delta)/H_{\mathcal{G}}(k_0, \Delta)$ is always larger than $\gamma(k_0, \Delta) > 0.95$. We conclude that a dichotomic position measurement is nearly optimal to estimate the height of the potential barrier Δ.

6. Conclusions

In this work, we introduce and discuss a general probing scheme for scattering problems based on continuous-time quantum walks. In particular, we consider a one-dimensional lattice, with an impurity at its center, i.e., a potential barrier of height Δ, and discuss in details how to quantify the maximum extractable information about the parameter Δ.

Using the continuous-space case as a guide for attacking the problem, we first introduce the single-momentum scattered states $S|k\rangle$ and use them to compute the reflection and transmission coefficients of the considered potential. From the scattered states, we build up the asymptotic Gaussian states, i.e., physical states, that depend, in addition to Δ, on the initial central momentum k_0 and the spread of the wave packet in position space σ. We then derive the reflection and transmission probability of such wave packets. Finally, we compute the QFI for the parameter Δ. We show that the QFI has a maximum for $\Delta = 0$, and it is strongly affected by the value of k_0. In particular, values of k_0 near $\frac{\pi}{2}$ lead to a smaller QFI. Moreover, for $|\Delta| \ll 1$, a small σ can increase the precision of the estimation. However, inspection of the QSNR did not show a noticeable difference in its behavior depending on the value of σ or k_0. The QSNR has a maximum for $\Delta \neq 0$, indicating that given the value of the central momentum k_0, there exists a value for Δ that can be better estimated, leading to the unit QSNR independently of σ and k_0.

Finally, we investigate the performances of a dichotomic position measurement, which is a binary measurement that is just able to distinguish if a particle is located on the left (reflected) or on the right (transmitted) of the potential barrier. We show that this measurement is optimal, i.e., its FI is equal to the QFI, for large initial wave packets (in position space), while it is nearly optimal for narrow initial wave packets.

Our work paves the way toward the characterization of more involved forms of potentials using a single-particle continuous-time quantum walk as a probe. Extensions of this work may also include more complex structures, such as multi-dimensional graphs, where imperfections created during the fabrication process need to be estimated in order to better control the quantum dynamics over such networks.

Author Contributions: Conceptualization, C.B. and M.G.A.P.; Formal analysis, F.Z., C.B. and M.G.A.P.; Investigation, F.Z.; Methodology, C.B. and M.G.A.P.; Software, F.Z.; Supervision, M.G.A.P.; Validation, C.B. and M.G.A.P.; Visualization, C.B.; Writing—original draft, F.Z. and C.B.; Writing—review & editing, C.B. and M.G.A.P. All authors have read and agreed to the published version of the manuscript.

Funding: This research received no external funding.

Acknowledgments: MGAP is a member of INdAM-GNFM.

Conflicts of Interest: The authors declare no conflict of interest.

Abbreviations

The following abbreviations are used in this manuscript:

CTQW   Continuous-time quantum walk
CR   Cramér–Rao
FI   Fisher information
QFI   Quantum Fisher information
QSNR   Quantum signal-to-noise ratio

Appendix A. Gaussian Wave Packet in K-Space

Consider the Gaussian wave packet in position space defined by Equation (21). Here, we show that its expression in k-space, within certain approximations, is given by Expression (23). We start by considering the the projection of Equation (21) into a state $|k\rangle$:

$$\langle k|\mathcal{G}_{k_0}\rangle = \frac{\mathcal{N}}{\sqrt{2\pi}} \sum_{j\in\mathbb{Z}} e^{-\frac{(j-\mu)^2}{2\sigma^2}} e^{i(k_0-k)j}. \tag{A1}$$

The infinite sum can be calculated using the Poisson summation formula, which states that, for suitable functions f: $\sum_{j\in\mathbb{Z}} f(j) = \sum_{n\in\mathbb{Z}} \hat{f}(n) = \sum_{n\in\mathbb{Z}} \int_{-\infty}^{+\infty} f(x)e^{-i2\pi nx}\,dx$. In our particular case:

$$\langle k|\mathcal{G}_{k_0}\rangle = \frac{\mathcal{N}}{\sqrt{2\pi}} \sum_{j\in\mathbb{Z}} e^{-\frac{(j-\mu)^2}{2\sigma^2}} e^{i(k_0-k)j} = \frac{\mathcal{N}}{\sqrt{2\pi}} \sum_{n\in\mathbb{Z}} \int_{-\infty}^{+\infty} e^{-\frac{(x-\mu)^2}{2\sigma^2}} e^{i(k_0-k)x} e^{-i2\pi nx}\,dx. \tag{A2}$$

The last integral is a continuous Fourier transform of a Gaussian function; therefore:

$$\int_{-\infty}^{+\infty} e^{-\frac{(x-\mu)^2}{2\sigma^2}} e^{i(k_0-k)x} e^{-i2\pi nx}\,dx = \sqrt{2\pi\sigma^2} e^{-\frac{(2\pi n+k-k_0)^2}{2\frac{1}{\sigma^2}}} e^{-i\mu(2\pi n+k-k_0)}. \tag{A3}$$

Inserting Equation (A3) into (A2) (discarding the constant global phase $e^{i\mu k_0}$), we obtain:

$$\langle k|\mathcal{G}_{k_0}\rangle = \mathcal{N}\sigma \sum_{n\in\mathbb{Z}} e^{-\frac{(2\pi n+k-k_0)^2}{2\frac{1}{\sigma^2}}} e^{-i\mu(2\pi n+k)}. \tag{A4}$$

The transformed state is not a Gaussian state, but it is an infinite sum of Gaussian states periodically displaced. However, if the wave packet is localized enough in reciprocal space, it is possible to approximate the last infinite summation by keeping only the central term $n = 0$ (it is always possible to shift the definition of k and k_0 in the interval $[-\pi, \pi)$ because they are defined modulo 2π). The localization assumption is needed in order to consider only one term, otherwise the tails of adjacent Gaussian functions could overlap. In other words, k_0 should be far away from the boundaries of the interval $[-\pi, \pi)$, i.e., $|k_0 - \pi|$ and $|k_0 + \pi|$ should be much larger than the standard deviation in the reciprocal space. Overall, the conditions read as $\min(|k_0 - \pi|, |k_0 + \pi|) \gg 1/\sigma$. With this assumption:

$$\mathcal{G}_{k_0}(k) = \langle k|\mathcal{G}_{k_0}\rangle \approx \mathcal{N}\sigma e^{-\frac{(k-k_0)^2}{2\frac{1}{\sigma^2}}} e^{-i\mu k}. \tag{A5}$$

Thus, a discrete Gaussian state in the position basis remains a Gaussian state in reciprocal space within the considered approximation. The calculation of the normalization constant $\mathcal{N}$ reduces to the calculation of a Gaussian integral:

$$1 = \int_{-\pi}^{\pi} \left| \mathcal{G}_{k_0}(k) \right|^2 dk \approx \int_{-\infty}^{\infty} \left| \mathcal{G}_{k_0}(k) \right|^2 dk = |\mathcal{N}|^2 \sigma \sqrt{\pi}, \tag{A6}$$

with:

$$|\mathcal{N}|^2 \approx \frac{1}{\sqrt{\pi \sigma^2}}. \tag{A7}$$

Appendix B. The Explicit Expression of the Functions $g_H(\Delta, k_0)$ and $g_F(\Delta, k_0)$

We have:

$$g_H(k_0, \Delta) = \frac{4 \left[3 \cos 6k_0 + 2(5\Delta^2 - 1) \cos 4k_0 + 3(3\Delta^4 - 19) \cos 2k_0 + \Delta^4 - 10\Delta^2 + 18 \right]}{\left[\Delta^2 + 2(1 - \cos 2k_0) \right]^4}, \tag{A8}$$

$$g_F(k_0, \Delta) = \frac{8 \left[\cos 6k_0 + 6\Delta^2 \cos 4k_0 + (\Delta^4 - 9) \cos 2k_0 - 6\Delta^2 + 8 \right]}{\left[\Delta^2 + 2(1 - \cos 2k_0) \right]^4}. \tag{A9}$$

References

1. Rutherford, E.F. LXXIX. The scattering of α and β particles by matter and the structure of the atom. *Philos. Mag. Ser.* **1911**, *21*, 669–688. [CrossRef]
2. Franklin, R. Influence of the Bonding Electrons on the Scattering of X-Rays by Carbon. *Nature* **1950**, *165*, 71–72. [CrossRef] [PubMed]
3. Chamberlain, O.; Segrè, E.; Wiegand, C.; Ypsilantis, T. Observation of Antiprotons. *Phys. Rev.* **1955**, *100*, 947–950. [CrossRef]
4. Aad, G.Et Al.. Obs. A New Part. Search Stand. Model Higgs Boson ATLAS Detect. LHC. *Phys. Lett. B* **2012**, *716*, 1–29. [CrossRef]
5. Helstrom, C.W. *Quantum Detection and Estimation Theory*; Academic Press: New York, NY, USA, 1976.
6. Gebbia, F.; Benedetti, C.; Benatti, F.; Floreanini, R.; Bina, M.; Paris, M.G.A. Two-qubit quantum probes for the temperature of an Ohmic environment. *Phys. Rev. A* **2020**, *101*, 032112. [CrossRef]
7. Tamascelli, D.; Benedetti, C.; Breuer, H.P.; Paris, M. Quantum probing beyond pure dephasing. *New J. Phys.* **2020**, *22*, 083027. [CrossRef]
8. Salari Sehdaran, F.; Bina, M.; Benedetti, C.; Paris, M. Quantum Probes for Ohmic Environments at Thermal Equilibrium. *Entropy* **2019**, *21*, 486. [CrossRef]
9. Mirkin, N.; Larocca, M.; Wisniacki, D. Quantum metrology in a non-Markovian quantum evolution. *Phys. Rev. A* **2020**, *102*, 022618. [CrossRef]
10. Wu, W.; Shi, C. Quantum parameter estimation in a dissipative environment. *Phys. Rev. A* **2020**, *102*, 032607. [CrossRef]
11. Seveso, L.; Benedetti, C.; Paris, M. The walker speaks its graph: Global and nearly-local probing of the tunnelling amplitude in continuous-time quantum walks. *J. Phys. A Math. Theor.* **2019**, *52*, 105304. [CrossRef]
12. Tamascelli, D.; Benedetti, C.; Olivares, S.; Paris, M.G.A. Characterization of qubit chains by Feynman probes. *Phys. Rev. A* **2016**, *94*, 042129. [CrossRef]
13. Razzoli, L.; Ghirardi, L.; Siloi, I.; Bordone, P.; Paris, M.G.A. Lattice quantum magnetometry. *Phys. Rev. A* **2019**, *99*, 062330. [CrossRef]

Entropy **2020**, *22*, 1321

14. Benedetti, C.; Paris, M. Characterization of classical Gaussian processes using quantum probes. *Phys. Lett. A* **2014**, *378*, 2495–2500. [CrossRef]

15. Schreiber, A.; Cassemiro, K.N.; Potoček, V.; Gábris, A.; Jex, I.; Silberhorn, C. Decoherence and Disorder in Quantum Walks: From Ballistic Spread to Localization. *Phys. Rev. Lett.* **2011**, *106*, 180403. [CrossRef]

16. Izaac, J.A.; Wang, J.B.; Li, Z.J. Continuous-time quantum walks with defects and disorder. *Phys. Rev. A* **2013**, *88*, 042334. [CrossRef]

17. Li, Z.J.; Izaac, J.A.; Wang, J.B. Position-defect-induced reflection, trapping, transmission, and resonance in quantum walks. *Phys. Rev. A* **2013**, *87*, 012314. [CrossRef]

18. Li, Z.; Wang, J. Single-point position and transition defects in continuous time quantum walks. *Sci. Rep.* **2015**, *5*, 13585. [CrossRef]

19. de Falco, D.; Tamascelli, D. Noise-assisted quantum transport and computation. *J. Phys. A Math. Theor.* **2013**, *46*, 225301. [CrossRef]

20. Chakraborty, S.; Novo, L.; Di Giorgio, S.; Omar, Y. Optimal Quantum Spatial Search on Random Temporal Networks. *Phys. Rev. Lett.* **2017**, *119*, 220503. [CrossRef]

21. Cattaneo, M.; Rossi, M.A.C.; Paris, M.G.A.; Maniscalco, S. Quantum spatial search on graphs subject to dynamical noise. *Phys. Rev. A* **2018**, *98*, 052347. [CrossRef]

22. Morley, J.G.; Chancellor, N.; Bose, S.; Kendon, V. Quantum search with hybrid adiabatic–quantum-walk algorithms and realistic noise. *Phys. Rev. A* **2019**, *99*, 022339. [CrossRef]

23. Benedetti, C.; Rossi, M.A.C.; Paris, M.G.A. Continuous-time quantum walks on dynamical percolation graphs. *EPL* **2019**, *124*, 60001. [CrossRef]

24. Boykin, T.B.; Klimeck, G. The discretized Schrödinger equation and simple models for semiconductor quantum wells. *Eur. J. Phys.* **2004**, *25*, 503–514. [CrossRef]

25. Tarasov, V.E. Exact discretization of Schrödinger equation. *Phys. Lett. A* **2016**, *380*, 68–75. [CrossRef]

26. Farhi, E.; Gutmann, S. Quantum computation and decision trees. *Phys. Rev. A* **1998**, *58*, 915–928. [CrossRef]

27. Mülken, O.; Blumen, A. Continuous-time quantum walks: Models for coherent transport on complex networks. *Phys. Rep.* **2011**, *502*, 37–87. [CrossRef]

28. Wong, T.G.; Tarrataca, L.; Nahimov, N. Laplacian versus adjacency matrix in quantum walk search. *Quantum Inf. Proc.* **2016**, *15*, 4029–4048. [CrossRef]

29. Simon, S.H. *The Oxford Solid State Basics*; The Oxford University Press: Oxford,UK, 2013.

30. Paris, M.G.A. Quantum estimation for quantum technology. *Int. J. Quantum Inf.* **2009**, *7*, 125–137. [CrossRef]

31. Griffiths, D.J. *Introduction to Quantum Mechanics*, 2nd ed.; Prentice Hall: Upper Saddle River, NJ, USA, 2005.

Article

Complex Systems in Phase Space

David K. Ferry [1,*], Mihail Nedjalkov [2,3], Josef Weinbub [4], Mauro Ballicchia [2], Ian Welland [1] and Siegfried Selberherr [2]

[1] School of Electrical, Computer, and Energy Engineering, Arizona State University, Tempe, AZ 25287-5706, USA; iwelland@asu.edu

[2] Institute for Microelectronics, TU Wien, 1040 Vienna, Austria; mihail.nedialkov@tuwien.ac.at (M.N.); mauro.ballicchia@tuwien.ac.at (M.B.); siegfried.selberherr@tuwien.ac.at (S.S.)

[3] Bulgarian Academy of Sciences, 1113 Sofia, Bulgaria

[4] Christian Doppler Laboratory for High Performance TCAD, Institute for Microelectronics, TU Wien, 1040 Vienna, Austria; josef.weinbub@tuwien.ac.at

*   Correspondence: ferry@asu.edu

Received: 22 August 2020; Accepted: 24 September 2020; Published: 29 September 2020

Abstract: The continued reduction of semiconductor device feature sizes towards the single-digit nanometer regime involves a variety of quantum effects. Modeling quantum effects in phase space in terms of the Wigner transport equation has evolved to be a very effective approach to describe such scaled down complex systems, accounting from full quantum processes to dissipation dominated transport regimes including transients. Here, we discuss the challanges, myths, and opportunities that arise in the study of these complex systems, and particularly the advantages of using phase space notions. The development of particle-based techniques for solving the transport equation and obtaining the Wigner function has led to efficient simulation approaches that couple well to the corresponding classical dynamics. One particular advantage is the ability to clearly illuminate the entanglement that can arise in the quantum system, thus allowing the direct observation of many quantum phenomena.

Keywords: nonlinearity; hysteresis; quantum transport; non-Hermitian behavior

---

1. Introduction

Over the past few decades, technological momentum has pushed semiconductors to the nanometer scale and has even led to structural modifications of the basic field-effect transistor (FET), such as replacing the planar FET with the finFET or Trigate FET using a vertical channel [1,2]. In such devices, the active region is restricted to nanometer-scale dimensions in one or more directions, which, depending on the involved time and energy scales, gives rise to quantum effects, such as energy quantization, tunneling, position-momentum uncertainty, phase coherence, etc., and these make the carrier transport quite complex. The complexity arises not merely from the need for a deeper understanding of transport and behavior of such small systems, but also because these small devices interact strongly with their environment. From a physical point of view, a small device comprises an active region, which is open to the environment in which it is embedded (and thus exchanges carriers is subject to interactions with this environment). This connection to the environment may be through a set of portals, described as contacts, or through interactions with the phonon structure of the lattice upon which the device lies, or through other types of interactions. The central feature of such small devices is that the device micro-dynamics cannot be treated in isolation and must be considered in conjunction with this environment [3]. This leads to a central tenet that the transport is now heavily influenced by this coupling to the environment, and the basic Liouville equation and its causal boundary conditions must be modified to account for the influence of this environment. In particular, the environment dramatically changes the quantum nature of the device, and the device similarly must have an effect upon the environment, as measurements can only be made in the environment [4]. In considering transport in these small systems, further complexity arises because it has lost its time reversible properties. A phase transition to irreversible behavior has occurred [5]. Hence, the device in particular is now a far-from-equilibrium, complex system.

There is an additional problem in small semiconductor devices, which further complicates the behavior. An analysis of the classical transport always gives us insight into the challenges that arise in quantum transport models. In small devices, the time scale of carrier transport, within the device, may well be dominated by the transient response characteristics of the carrier velocity and distribution function [6,7]. Then, there may be excitation/relaxation effects in the environment, which affect the transient behavior of the device itself through the device-environment interactions. Each of these effects provides considerable complications and sets requirements upon any approach to quantum transport to be applied in such nano-devices. This is not the least because the transient excitation response is usually quite different from the relaxation response. As may be expected, this further complicates the far-from-equilibrium treatment of the device.

When we drive a semiconductor system out of equilibrium, the resulting distribution function used to characterize the transport does not simply evolve from the equilibrium version. Rather, it evolves into a balance between the driving forces and the dissipative forces (assuming that a steady-state balance can be achieved). There is a hierarchy of equations used to determine this distribution function, as we move from large classical systems to smaller fully quantum mechanical systems, and, in fact, there is a hierarchy of quantum mechanical approaches. The most detailed classical approach uses the Boltzmann transport equation, and this transitions to several "Boltzmann-like" quantum analogs, which arise from the density matrix, the Wigner function and non-equilibrium Green's functions as one moves down the hierarchy. Much has been written about quantum transport, especially with regard to semiconductor devices. Unfortunately, a great deal of this material has not taken proper account of the far-from-equilibrium behavior that these devices exhibit, the difficulties of short-time response, or the complicated interactions between the environment and the device.

Our purpose in this article is not to review this entire body of work, but rather to try to illustrate the nature of the problems that face someone trying to make sense of the complex system with which one desires to work. In the next section, we will discuss some of the attributes of complex systems, whether classical or quantum, as well as how classical and quantum systems differ. In Section 3, we will explore the difficulties that arise due to the arrow of time and the resulting irreversibility. Then, in the following sections, we discuss the leading methods of quantum transport and their advantages and disadvantages. This will mention the nonequilibrium Green's functions but mainly rely upon phase-space Wigner functions. We will actually do these last two in the reverse order, because we believe the former is limited in complex systems, while the latter allows us to utilize numerical tools—the ensemble Monte Carlo particle methods—that are directly transferred from classical transport. We will illustrate the methods with new results from four different applications, some of which have never been studied with Monte Carlo methods previously.

2. The Nature of Complex Quantum Systems

A complex system may be regarded to be any system that is composed of multiple parts, many of which are interacting with one another, which fits the above description of the device and its environment perfectly. This assembly of various parts are typically both nonlinear and inhomogeneous, especially when we also consider the environment in which the system is embedded. For our purposes, we can thus regard any electronic nano-device as being a complex system, as this device certainly interacts with its environment in a way that changes the properties of both the device and the environment, a point we deal with below in detail.

To begin, let us consider the system itself without its environment. We know that the set of energy levels for a quantum object allow us to express the Hamiltonian in terms of the extracted eigen-states for these levels; this is the energy eigen-state description and the Hamiltonian is diagonal in the absence of interactions, as

$$H_0\varphi_n(x,t) = E_{0n}\varphi_n(x,t), \tag{1}$$

where the $\{\varphi_n(x,t)\}$ form a complete set of basis functions as mentioned. For interactions within the system, such as with scattering as an interaction, we treat the time-dependent perturbation with the Fermi-golden rule. But, this does not conserve normalization of the state. To conserve normalization, we have to use self-consistency, which in this case means accounting for the decay of the initial state [8]. When this is done, one determines a self-energy given as $\Sigma_n = \Delta_n + i\Gamma_n$. The real part of the self-energy corresponds to a shift downward of the energy (frequency shift due to the energy shift $E_{0n} \rightarrow E_{0n} - \Sigma_n$). The imaginary part of the self-energy provides a damping of the state in time, which is the analog of the classical resistive damping of a resonant circuit. The diagonal energies in the Hamiltonian have this self-energy correction for each energy level (as indicated, Σ may be different for each energy level), and the Hamiltonian has become non-Hermitian. In particular, an arrow of time has entered the description of our quantum object. More importantly, the interaction leads to the view that states which do not lie on the energy shell (described by the classical delta function between energy and its evaluation in terms of momentum) can be important in the transport. So, there is no longer any single energy shell, but a range of momentum values that can have the same energy. To describe this, one defines a new quantity, which is called the spectral density, and describes the relationship between the energy and the momentum, typically a Lorentzian line in equilibrium systems, just as in the classical case. We will deal with the spectral density further below.

Let us now embed our quantum object, the device, within a surrounding environment as discussed above. We can describe the states of the device by a density matrix ρ_D and the environment by a density matrix ρ_E, which form a tensor product when the two parts are uncoupled. Our goal is to see how they occur after the coupling and the interaction. The complex system can now be defined by a Hamiltonian containing three terms:

$$H = H_D + H_E + H_{int} \tag{2}$$

where H_D describes the device, H_E describes the environment, and H_{int} describes interactions between the two. The composite density matrix for the entire system begins with the tensor product of the two density matrices described above. There are two crucial steps in defining a reduced density matrix for just the desired parts of the device. The first is to project out these states via a projection super-operator. The second is to perform a trace over the environmental states which yields just the reduced set of pointer states. This procedure has been known for a considerable time [9–11], and the derivations of the following form have been discussed extensively [3,12,13]. The result is the transport equation for the projected/desired part of the device

$$i\hbar\frac{\partial \rho_{PD}}{\partial t} = \left(H_{PD} + H'_{int}\right)\rho_{PD} + \Sigma\rho_{PD} \tag{3}$$

where

$$\begin{aligned} H'_{int}\rho_{PD} &= Tr_E\{PH_{int}P\rho_D\} \\ \Sigma\rho_{PD} &= Tr_E\{C\rho_{PD}\} \\ C &= \tfrac{i}{\hbar}PHQe^{-iQHQt/\hbar}QHP \end{aligned} \tag{4}$$

and P and $Q = 1 - P$ are the projection super-operators that project the desired dynamics of the device onto a reduced density matrix, or to its conjugate parts, respectively. Once again, the net Hamiltonian is non-Hermitian. Moreover, the response of (3) is retarded, as is the usual case in projected systems. More importantly, the second term in the parentheses of (3) may contain new processes that are not part of either the device or the environment alone. Writing the second term in parentheses as separated, as in (3), is a short-hand notation, since ρ_D cannot actually be separated from this interaction term. In that sense, this term is an entanglement between environment and device, except that "entanglement" is also not a good description as this connection can appear even in classical systems (entanglement is usually reserved for quantum systems). This term can be new processes depending upon both the environment and the system. One such type is the resonant back-scattering trajectories from quantum dots in a magnetic field [14–16], which depends critically upon the actual confinement structure of the device and its contacts.

However, how does ρ_D differ from $\rho_{PD} = P\rho_D$? The former density matrix contains all the eigen-states of the entire device, given in (1), while the latter contains only those which can be used e.g., for modeling and simulation of the entire system. The separation arises from the coupling to the environment, which may well wash out a number of the quantum states. Those states which are not washed out, termed the pointer states, provide the quantum effects within the device. In a bulk semiconductor, this might just be the carrier dynamics of spatially quantized electrons (or holes). For the device, it at least contains all eigen-states which are necessary to describe the response of the device to excitations. For a counter example, the device contains the oxide, but the oxide polar modes are usually not seen in the response directly, yet they provide intra-device scattering processes as does the interface between the oxide and the semiconductor. These scattering processes are included above in C, as are those processes arising from environmental effects, such as lattice phonons. The form of C is standard perturbation theory where PHQ and QHP are versions of the matrix element coupling the device to the phonons and its adjoint, respectively, while the exponential contributes to the energy conservation through the appropriate frequencies. We now find that quantum transport, like classical transport has distinctly far-from-equilibrium behavior which can lead to non-Hermitian Hamiltonians and very nonlinear, inhomogeneous, and retarded transport, with new phenomena that are not present in equilibrium systems.

2.1. Environmentally-Induced States

Classically, a well known environmentally-induced effect in the device world is drain-induced barrier lowering [17]. Current injection into the MOS channel is governed by a potential barrier at the source end of the channel. The barrier height is controlled by the gate-source potential difference. Normally, there is enough scattering in the channel of the injected carriers, that the drain voltage is "screened" from this barrier. But, if ballistic transport begins to occur as the size of the transistor is reduced, the screening effect of the scattering is also significantly reduced. Then, the potential barrier begins to be affected by the drain potential which, in turn, can act to lower the barrier itself, letting more current into the channel [18]. We consider this as an environmentally-induced state as the drain voltage is an environmental variable whose effect is transmitted through one of the contacts of the device structure. The contacts themselves are a complex object whose properties are often not well behaved. Hence, the device does not actually see the real drain voltage, but only some complex image of it as transmitted into the actual device.

Now, let's turn to a quantum interaction. We pointed out above that the coupling between the environment and the device can lead to new states which do not exist in the device alone. We can illustrate this with an array of quantum dots which are open to each other and to the environment. Any two dots are coupled to each other via an open (non-tunneling) point contact-like structure, and the end dots of the array are coupled similarly to the environment. It has been found that such a structure supports new states that are localized on the quantum point contacts [14]. These states are stable states and contribute strongly to the overall conductance through the device, and have been shown to support the concept of quantum Darwinism [19], in which the exact same wave function amplitude is seen in each mode that propagates through the quantum point contact.

2.2. Trajectories and Quantum Mechanics

Trajectories have been used in classical mechanics for centuries, where a particle trajectory follows Hamilton's equations of motion. And, this has been extended to transport theory through a particle based simulation of the Boltzmann transport equation, which we normally call a kinetic Monte Carlo process [20]. But, shortly after the appearance of the Schrödinger equation [21], trajectories were suggested for quantum mechanics as well. Physical observables retain the same appearance as their classical counterparts, following the contributions of Madelung [22] and Kennard [23], who pointed out the quantum dynamics would follow the classical potential plus any quantum potential.

In several simulations, we have compared classical trajectories with the full quantum mechanical solutions [24–26]. In ballistic cases, the pointer states in classical simulations are located around singularities called centers, as the eigenstate sits on a closed ring located at this point. The full quantum simulation is, of course, smoothed out, due to uncertainty. Nevertheless, a Husimi function (which is a smoothed Wigner function) for the quantum solution is located over the center corresponding to the classical orbit. In the case of the environmentally-induced state, we term this classical state as a bipartite state that is associated with a ring of attractors located in the quantum point contact region, and the quantum state projects onto this same phase-space region. Most of these simulations actually use an iterated form of the Schrödinger equation to obtain the quantum wave functions.

Over the past few decades, it has been realized that the Wigner function is particularly suited to simulation with an ensemble of particles through the Monte Carlo procedure [27–29]. This is because of the strong connection between the Wigner transport equation and the Boltzmann transport equation, discussed below. A particle model to be evaluated with a Monte Carlo technique has been associated with the Wigner and the Wigner–Boltzmann equation. This model makes the analogy between classical and Wigner transport formalisms even closer, but is certainly in keeping with the approach of Kennard. We will develop this more fully in Section 4 below.

3. Time Irreversibility

It was pointed out above that, when we apply fields or forces to the device, the entire complex system undergoes a phase transition that breaks time-reversal symmetry. The device, within its environment, then seeks a steady-state, far-from-equilibrium stable state that balances the driving forces and the dissipative forces. During the transient response toward this stable state, if it exists, the system may evolve though a number of intermediate phases, e.g., homogeneous, inhomogeneous, linear, nonlinear, etc. When the forces are removed, the system response does not reverse its course through these different phases, but seeks a relaxation toward the equilibrium steady-state from which it initially deviated. The excitation process generates entropy [30]. The relaxation does not remove this entropy from the system but generates even more as dissipation still occurs. We briefly will describe such a case and then consider another, more complicated example.

To illustrate the difference between excitation and relaxation, consider the so-called Gunn diode, composed of a bulk GaAs device. Typically, it is moderately doped in the 10^{14}–10^{16} cm^{-3} region. This device is an example of a negative-differential conductance (NDC) device [31]. The typical current-voltage curve is illustrated in Figure 1. Here, there are regions of the curve where a single current density can be supported by multiple values of electric field (see the dashed line). Hence, we think of this curve as $J(E)$, and thus of conductance. Thus, the use of NDC. It is important to note that, for the dashed line, the central crossing of the $J(E)$ curve, where J is decreasing while E is increasing, is unstable and certainly is not a position of steady state.

When the device of Figure 1 is excited, the current increases along the blue curve until the peak is reached in the current. Then, the operating point will try to jump to the high electric field region. But the path it takes will not follow the blue curve, but one closer to the red curve marked "excitation". It may even move to a much higher current if the switching is fast, as any overshoot will appear as a higher current. So, the exact red path followed depends upon how fast the device is driven to a high-electric field. If the device reaches a steady-state, perhaps with a current between the peak and the valley, an inhomogeneous electric field and charge density will exist in the device (the horizontal dashed line in Figure 1 represents an average current for the inhomogeneous electric field). There will be a high-electric-field region and a low-electric-field region, and an inhomogeneous charge density is required to support this field difference through Gauss' law. Now, it may turn out that it is not possible to reach a steady state, for the reason that the carriers are moving with the drift velocity appropriate for the dashed curve. As the carriers reach the drain and exit the device, the field will try to become homogeneous once again and this will trigger another high-field region near the cathode. This then propagates through the device until it reaches the anode and starts the process over again. This leads to current (and voltage) oscillations first observed by Gunn [32]. When the voltage excitation is removed, the operating point does not pass back over the peak current density, but follows the lower red arrow marked "relaxation" through the valley current density, then jumping back to the low-field current. This leads to a hysteresis in the current-voltage characteristics. When the device and its environment are considered, the nonlinear hysteretic behavior is considered to be an elegant example of catastrophe theory [33], in which the effective potential generates a so-called fold catastrophe.

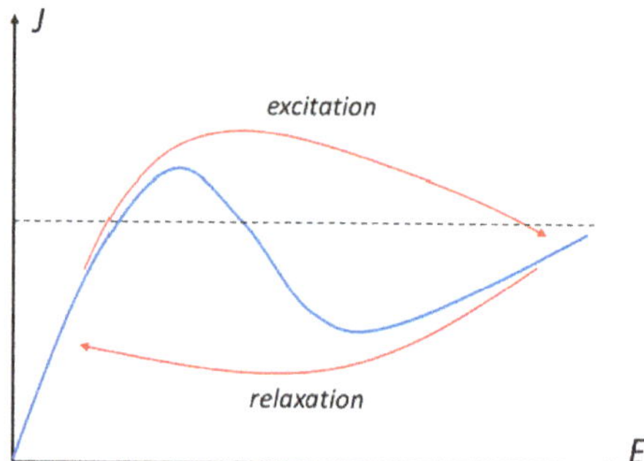

Figure 1. The current-field curve for an NDC device (blue curve). The red arrows indicate the differences between excitation and relaxation, explained further in the text.

We conclude that carrier transport in complex quantum systems, such as nano-scale semiconductor devices embedded within a complicated environment, requires a complex quantum description, which takes into account quantum-coherent phenomena, as well as dissipative processes of scattering, with both modified by the interaction with the environment. In the following, we will discuss the physically intuitive Wigner phase-space function [34–36]. Our preferred focus is upon this function, mainly because of its adaptability to ensemble Monte Carlo (EMC) simulations, and we will finally describe a number of examples of this approach.

4. Wigner Functions

In general, we seek a formulation of many-body statistical quantum mechanics using the methods of the interaction representation, perturbation theory, and second quantization in terms of expectation values of field operators. To begin, we need to clarify just what this last sentence means in practice. In Section 2, we broadly introduced the density matrix. The usefulness of this lies in the existence of an entire set of basis functions, which are characteristic of the problem at hand. These lead to a density matrix as

$$\rho(x, x', t) = \sum_{nm} c_{nm} \varphi_m^\dagger(x', t) \varphi_n(x, t). \tag{5}$$

Normally, the trace of the density matrix, which is the magnitude squared of the basis functions in the absence of interactions, gives the average of a physical quantity as

$$\langle A \rangle = Tr\{A\rho\}. \tag{6}$$

Temporal evolution of the density matrix is governed by the Liouville equation

$$i\hbar \frac{\partial \rho}{\partial t} = \left[-\frac{\hbar^2}{2m} \left(\frac{\partial^2}{\partial x^2} - \frac{\partial^2}{\partial x'^2} \right) + V(x) - V(x') \right] \rho. \tag{7}$$

From this equation, one can develop an effective Boltzmann-like equation of motion, as is done below. As with the wave function approach above, the density matrix is easy to couple to the self-consistent Poisson equation for devices.

The Wigner function formalism presents a physically intuitive formulation of the quantum mechanical theory, which retains most of the classical concepts and notions. The Wigner function, which is defined in the phase space, is the quantum mechanical analog of the classical distribution function. It is a real-valued function, often called a quasi-distribution, since physical averages are obtained by classical, distribution function-based expressions. However, it allows for negative values as a result of the uncertainty relation and quantum information/entanglement. The development of the Wigner phase-space formulation of quantum mechanics, and the appropriate mathematics for operators in this formulation, has been established historically [37]. This was followed by important contributions from the work of Groenewold [38] and Moyal [39]. Importantly, the Wigner function formalism has been established as an equivalent, autonomous alternative to operator mechanics; for a recent review see [35]. A self-contained formulation of the quantum theory in terms of phase-space functions has been attained by rules for filtering the admissible quantum states from the c-number functions of the phase space. Operators in this phase-space formulation are related by novel non-commutative algebraic rules given by Moyal. Important questions, like what discriminates classical from quantum mechanical behavior in the phase space have been addressed. Indeed, in equilibrium, the Wigner function is positive definite, which essentially couples it to the Boltzmann distribution function. However, the onset of quantum effects leads to both negative excursions and non-Gaussian distribution functions. Furthermore, it has been shown that phase-space quantum mechanics recovers the operator mechanics, so that it is clear that there is a logical equivalence between the two theories [40].

We introduce the Wigner function in the historical manner, beginning with the mixed state, single-time density matrix (5) and the Liouville equation governing it's evolution (7). Then, we introduce the center-of-mass coordinates for position,

$$R = \tfrac{1}{2}(x + x') , \quad s = x - x'. \tag{8}$$

and construct the Fourier–Weyl transform in the difference variable as

$$f_W(R, p, t) = \frac{1}{2\pi\hbar} \int_{-\infty}^{\infty} \rho\left(R + \frac{s}{2}, R - \frac{s}{2}\right) e^{-isp/\hbar}. \tag{9}$$

This leads to the Wigner transport equation as

$$\frac{\partial f_W}{\partial t} + \frac{p}{m}\frac{\partial f_W}{\partial R} = \frac{1}{2\pi\hbar} \int dp' V_W(R, p - p') f_W(R, p', t), \tag{10}$$

where the Wigner potential is given by

$$V_W(R, p) = \int ds \, sin\left(\frac{p's}{\hbar}\right)\left[V\left(R + \frac{s}{2}\right) - V\left(R - \frac{s}{2}\right)\right]. \tag{11}$$

The Wigner function is integrable with respect to position or momentum, and these integrals give rise to momentum or spatial carrier density distributions, respectively. The Wigner transport Equation (10) reveals an important analogy with the classical Boltzmann equation. The left hand side represents the Liouville free-streaming operator. We recognize that the Wigner function is a quasi-distribution, but it is sufficient to determine transport coefficients, given (10). That is, the Wigner function plays the same role as the Maxwell–Boltzmann distribution classically; once the Wigner function is known, all transport properties are determined from appropriate integrals over the distribution.

If the potential V in (11) is a linear or quadratic function, the right-hand side reduces to the classical force acceleration term so that (10) reduces to the ballistic Boltzmann equation. If the potential is a slowly varying function in the region where fw is sizable (characterized by a quantity called coherence length), the latter approaches the classical distribution function. Thus, the first important property is that the Wigner formalism ensures a seamless transition from quantum mechanical to classical transport regimes. This also reveals the basic difference between classical and quantum mechanical evolution in the phase space: The former (classical evolution) is governed by the normal force, corresponding to the first derivative of the Wigner potential, while all potential derivatives are involved in the latter (quantum evolution). Indeed, Wigner himself proposed that the leading term in the quantum correction would be given by the second derivative of the total potential.

Furthermore, a scattering operator, analogous to the one used with the classical Boltzmann equation, acting on fw may be added to the right-hand side of (10), giving rise to the Wigner-Boltzmann equation [41]. This equation describes the electron evolution as a competition between the scales of the involved physical quantities, e.g., the competition between the accelerating forces and the dissipative forces. If one develops dimensionless parameters, corresponding to the relative strength of the energy scales of the device, the potential, the phonon energy, and the electron–phonon coupling factor, these are tied into a recently derived notion called the scaling theorem [42]. This theorem reveals a second mechanism causing a seamless transition from quantum mechanical to classical transport. Scattering causes a coherence length reduction and electron localization. An increase in the electron-phonon coupling factor shrinks the spatial interval where the electronic state remains unperturbed or ballistic. This gradually transforms, on a microscopic level, the Wigner-Boltzmann equation into the classical Boltzmann counterpart. The second important property of the Wigner function formalism is that quantum-coherent and scattering-dominated transport regimes are treated on equal footing.

4.1. Particle Approaches

We already noted above that there is a view of quantum mechanics that specifically incorporates particle trajectories. Madelung took careful note of Schrödinger's work, and immediately noted that the probability density had all the appearances of a fluid flow. Kennard quickly learned of the new developments in quantum mechanics as well, and found the quantum mechanics of a system of particles came directly from the Schrödinger equation. Moreover, he found that the particles would follow normal Hamiltonian dynamics, although the potential would have to be modified through the addition of a quantum potential. One form of this potential is often called the Bohm potential following his resurrection of the Madelung–Kennard hydrodynamic ideas [43]. Further, of course, the quantum potential in which we are interested is the Wigner potential (11). Particles naturally move in phase space, so the Wigner phase-space representation of quantum mechanics is a natural venue for Monte Carlo particle dynamics. The key problem is just how to handle the complicated integral of the Wigner potential that appears in the Wigner transport Equation (10). There have been many methods developed, but we focus here on the pseudo-particle approaches that are used in Monte Carlo.

The weighted Monte Carlo approach was formally introduced in 1992 [44]. The paths considered have the problem that there are regions in phase space in which the distribution function has little weight, or is negative. This means that the particles in the Monte Carlo simulation have low probabilities of reaching these areas and the consequent solutions are very noisy in these phase space areas. In some sense, this approach arises from older ideas of variations in importance sampling to reduce variances in Monte Carlo. The weighted Monte Carlo procedure has been shown to be especially useful in the backward Monte Carlo, where one uses the series expansion to guide a propagation path from the final state back to the initial state. The path integral is expanded in a Neumann series as above [45,46]. The use of this series then corresponds to a set of paths that contain a different number of scattering interaction events. Since the paths involve a number of scattering events, each scattering event introduces a phase proportional to

$$2\cos\left[q(x_1 - x_2) - \omega_q(t_1 - t_2)\right],\tag{12}$$

where q and ω_q are the wave number and radian frequency of the phonon. Hence, because the scattering is nonlocal, it goes beyond the Fermi golden rule used in Boltzmann transport. Presumably, after many scattering events, multiple paths can be summed to provide the negative regions of the Wigner function. Now, we note that there are two positions and two times required to formulate the scattering process, and this is a complication for the method. Moreover, here the negative regions arise solely from the scattering processes.

The affinity method is another approach where an affinity, which may be negative, is assigned to each particle. This immediately solves the problem discussed in the last paragraph, since a negative part of the Wigner function is clearly represented by particles whose affinity is also negative. The Monte Carlo approach is set up so that two systems are solved simultaneously. The first system is the particle system, which resembles a standard classical EMC. The second system is the wave properties of the particles, the affinity. That is, all particles in the system are treated classically as whole particles, but the method accounts for fully coherent transport, but has been further generalized to account for scattering. They are scattered using normal EMC scattering techniques, and are drifted and accelerated using the standard field term deriving from the solution of the Poisson equation in the presence of the real potential, such as the tunneling barriers. However, the discontinuities in the potential are handled as boundary conditions on the solutions of the Poisson equation. That is, the potential jump is introduced in matching solutions from different regions of the solution space for the Poisson equation. Once the above operations have completed, the Wigner distribution function is calculated from the particle's position and affinity according to [47–49]

$$f_W(x,p) \sim \sum_i \alpha_i \delta(x - x_i)\delta(p - p_i),\tag{13}$$

where p_i, x_i, and α_i are the momentum, position and affinity, respectively, and the sum i runs over the set of particles used in the simulation. There are two points here. By using this in the Wigner equation of motion, we see that one needs to update both the classical properties (position, momentum, etc.) and the quantum properties (affinity). While the classical properties are updated by the classical forces, the affinity is updated by the quantum, or Wigner, potential. Because the collisions are classical, they have to be modified for the two times that occur in (12). Actual collision durations tend to be a few femtosecond [50]. Once we have a distribution for these times, then the nonlocality of the scattering can be easily introduced for each scattering event [51].

Another approach is the signed particle approach, in which the action of the Wigner potential gives rise to generation and annihilation processes: Any particle, with sign $a = \pm 1$, and momentum p, generates two new secondary particles, with signs $\pm a$ and momenta $(p \pm p_1)$. Both the generation rate and the distribution of momentum offsets p_1 are determined by the Wigner potential. Any of the three particles can again generate new, secondary pairs, etc. An important property of the model is that two particles with opposite sign, located in the same phase space point at the same time, annihilate each other. However, in numerical applications, the phase space has to be decomposed into cells rather than points, where the annihilation property gives rise to the concept of indistinguishable particles, which are stored in cells at consecutive time steps, so that a single integer number per cell replaces the ensemble of particle states within that cell. This greatly reduces the memory requirements in the implementation of the model. The first two-dimensional Wigner function based simulations could only be realized thanks to this approach but also due to novel parallelization strategies [52,53]. Because of both accomplishments, the number of particles can be significantly increased while still retaining feasible simulation times.

A variety of applications has been explored using the particle interpretation of the Wigner function formalism. As already discussed, a comparison between NEGF and Wigner simulations of RTD has been presented [54], showing how the Wigner function based approach bridges the gap between purely coherent and scattering dominated transport regimes. In the next section, we give some further illustrative examples of the Wigner Monte Carlo method.

4.2. Why Not Non-Equilibrium Green's Functions (NEGF)?

The NEGF has been widely applied to simulations of semiconductor devices, despite concerns over the applicability of many of the approaches. Some of the more advanced approaches use atomistic approaches to yield the full band structure. This has allowed the use of realistic band structures [55–57], phonon spectra [58], strain effects [59], interface roughness [60], and material characteristics. For purely ballistic problems, the NEGF formalism is the approach of choice, as it provides a tomography of the generic physical quantities in terms of energy, although in this situation the problem reduces to merely the retarded and advanced equilibrium Green's functions [61]. In general, the NEGF approach is efficient for stationary problems—determined by the boundary conditions—near the coherent limit. This is because the center-of-mass coordinates (8) must also be generated in the two-time coordinates. In nearly all work on NEGF, the average position R and the corresponding average time are totally ignored. Hence, these NEGF are for homogeneous steady-state systems. They do not, and often cannot, handle strongly inhomogeneous and transient systems.

There are further problems. Most modern approaches to the use of NEGF base their work on that of Keldysh [62]. The scattering self-energies depend on the carrier $G^<$ and $G^>$ and on the greater and lesser Green's functions of the phonons, which account for the occupancy of the phonon states and depend on the phonon energies. While this approach has been widely used, Keldysh points out that it works only upon the assumption that the system is close to equilibrium, so that the normal techniques can be used. For example, Keldysh assumes that the interaction representation, in which the scattering-derived interaction representation is assumed to be a unitary translation operator. But, this assumption fails when the energies contain self-energies so that the Hamiltonian, for example, (2), is no longer Hermitian. Again, most approaches for devices use only the Dyson equation [63] to provide a re-summation of interaction terms into a simple self-energy. With anisotropic scattering, such as that by impurities or polar phonons, one needs to utilize the Bethe–Salpeter equation [64] to completely evaluate the scattering. The onset of nonequilibrium phonon distributions [65], proper treatment of the scattering in many cases, and full use of both the Bethe–Salpeter equation with vertex corrections, increase tremendously the method's computational effort. It is then easy to see that particle-based Wigner functions allow a quicker route to the solution, as well as being more applicable than NEGF in many cases, particularly when inhomogeneous and transient transport is desired.

5. Wigner Function Applications

The Wigner function has been applied to a great many fields of science, so that there are a great many possible examples that could be discussed. Here, we will concentrate on four examples that have been investigated by the authors in a great many studies, and with new results. The first deals with electron tunneling, which has been discussed and treated with Wigner functions for some half a century, but it still admits to new insights. The second example comes from the study of spin transport in semiconductors, in which one common observable is the spin Hall effect, treated here for the first time by Monte Carlo techniques. Our third study deals with Josephson junction circuits, which are of interest to the quantum information and computing world, again treated for the first time by Monte Carlo techniques. Finally, we deal in the fourth example with electron interference in which local potentials, such as those arising from impurities, affect the quantum nature of scattering.

5.1. Tunneling

In Figure 2, we illustrate just how Wigner functions generate entanglement merely by passing through a tunneling junctions. Here, the simulation is based on the signed-particle model, and can be conveniently understood in classical terms thus enabling a deep physical insight about the evolution process. An initial carrier state encounters a potential with a specifically engineered shape, splitting the initial packet to four well established density peaks. The evolution maintains the initial coherence despite the fact that the peaks propagate in disparate directions. If time is reversed, the backward evolution recovers the initial state. This means that if one of these peaks is modified by another potential, the interaction will be felt by the whole state so that the rest of the peaks are modified too. Any such interaction causes a momentum redistribution of the signed particles representing the carrier state, and that is how we can establish an intuitive picture in which the peaks communicate via the momentum. This picture remains valid as long as the involved physical processes maintain the initial coherence. Phonons strive to redistribute the momentum of the signed particles towards equilibrium [66]. Then, the evolution loses it's quantum mechanical character in favor of the classical behavior, where an initial carrier ensemble is scattered by the potential, which modifies the probability distribution of the carriers in the phase space. These considerations illustrate processes that are important in, e.g., the area of quantum cryptography, where the primary obstacles are decoherence processes.

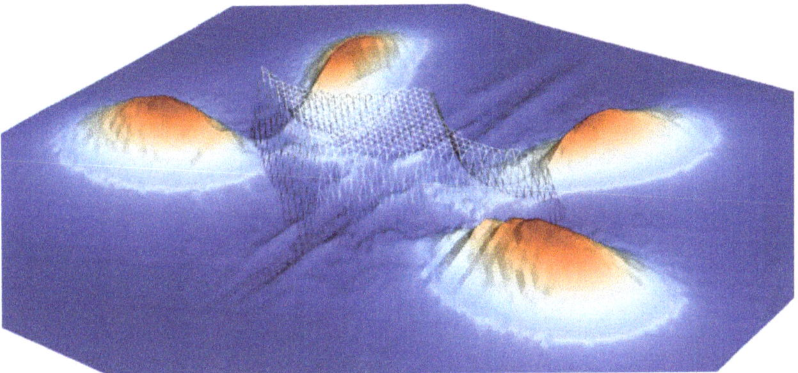

Figure 2. An initial Wigner function corresponding to a minimum uncertainty wave packet encounters a potential with a specifically engineered shape, splitting the initial packet into four well established density peaks propagating in disparate directions. The evolution maintains the initial coherence. The variations of the density in the potential region are related to the oscillations of the Wigner function, which furthermore connote interference effects and entanglement.

5.2. Spin Filtering

Gaps that open in the free electron spectrum in semiconductors typically give band energies of the Einstein (relativistic) form. This is especially true when the spin-orbit interaction is included. In the layer compounds, and particularly in the transition metal di-chalcogenides (TMDCs), in the presence of the spin-orbit interaction, the Hamiltonian can be written as [67]

$$H = at\left(\tau k_x \sigma_x + k_y \sigma_y\right) - \frac{\lambda \tau}{2}(\sigma_z - 1)s_z, \tag{14}$$

where the various σ terms are the Pauli matrices for two pseudo-spin basis functions of the valleys in the strange bands of the TMDC, τ is the valley pseudo-spin index, a is the lattice constant, t is the nearest neighbor hopping energy, Δ is the energy gap, 2λ is the spin splitting at the valence band top, and s_z is the Pauli matrix for spin. These materials lack an inversion symmetry, and the principal valence and conduction bands around the band gap derive primarily from the transition metal d-states [68], and they tend to have a direct band-gap at the K and K' points (corners of the hexagonal shape) of the Brillouin zone. The spin–orbit interaction produces opposite spin splittings in the two equivalent valleys of the valence band, and can lead to a valley-spin Hall effect, similar to the usual spin Hall effect [69–71], arising from the presence of a Berry curvature [72]. The spin is coupled to the valley pseudo-spin due to the difference in the orientation of the spin splitting [73]. This leads to a transverse velocity that is different in the two valleys and pushes the opposite spins to opposite sides of the nanowire.

In light of the above, it is clear that if we represent the particle by a Gaussian wave packet, it will move with a constant drift velocity that has a longitudinal component due to the applied electric field and a transverse component due to the Berry curvature and effective magnetic field. The latter will be oppositely directed for the two valleys. To illustrate this, we take the TMDC WS_2. Because the mobility of WS_2 is only of the order of 10–100 cm^2/Vs, the transverse velocity is not that much smaller than the longitudinal velocity. For a longitudinal field of 1 kV/cm, a typical density of 10^{11} cm^{-2} of free carriers, a mobility is 60 cm^2/Vs, we find a drift velocity of about 3.6×10^4 cm/s. The transverse velocity is generated by the Lorentz force arising from the effective magnetic field, and this gives a transverse velocity of approximately 1.9×10^4 cm/s, using values in (15) from [74]. We will use these values below in the simulations. For this system, the Wigner function for two spins starting from the same location in the semiconductor will be entangled with a cross-term representing the correlation of the two oppositely-directed spins [74], and this may be evaluated analytically. Using the particle Monte Carlo approach, we can continue to use a simple model for the Wigner function, formed initially from two Gaussian packets as in the analytical approach. However, we utilize a new sampling technique [75] for the Wigner potential to evaluate the entanglement of the two packets. In this approach, the wave-packet phase-space Monte Carlo method expands the wave-function in a local basis set (e.g., a Gaussian), then acquires via Ehrenfest's theorem a system of ordinary differential equations similar to the classical Hamilton's equations. Expectation values are then propagated in a manner similar to classical statistical mechanics. An interference pattern concentrated between the two arises from incorporating a phase related to the differences in real space and momentum space between the two packets. Particles are accelerated by using Ehrenfest's theorem for a real and quantum potential, whose value is found from the parameters of a parabolic band model of WS_2, and then utilizing the Hamiltonian Monte Carlo technique for solving the Wigner equation. The results are shown in Figure 3, and agree almost exactly with the analytical results. One may compare these results with those of [74] to observe that the Monte Carlo procedure faithfully reproduces the analytical results.

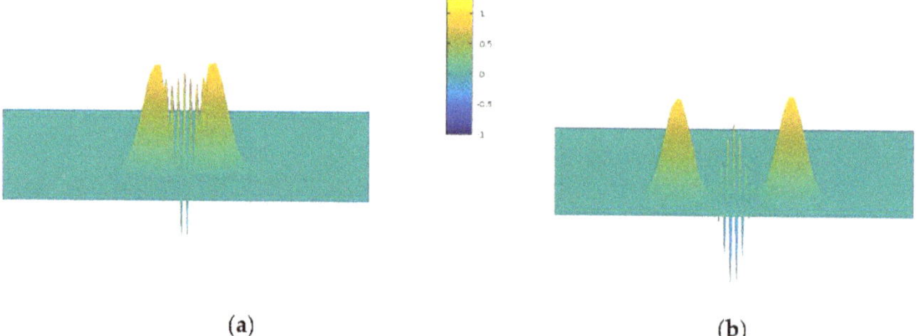

(a)   (b)

Figure 3. Wigner function for a pair of opposite spin electrons propagating via the valley spin Hall effect. The entanglement is clearly shown between the two main Gaussians. (**a**) shortly after the two spins states separate, where the entanglement still overlaps the main pulses. (**b**) At a later time, when the two main pulses are further separated and the entanglement is more distinct.

5.3. Superconducting Josephson Junction

The previous examples are known to be strongly connected to classical phase space solutions. In many cases, the equation simply becomes the Boltzmann equation. The characteristics of the Liouville operator are simply the classical trajectories. Now, we want to turn to non-quadratic Hamiltonians. The only nonlinear Hamiltonian known in quantum mechanics to have an analytical solution is the hydrogen atom. However, it would be of greater interest to see if the method can actually identify the eigenvalues and eigenfunctions for a Hamiltonian with unknown solutions, or at least not well-known solutions. The simplest case to examine is the tilted washboard, given by

$$H(x,p) = \frac{p^2}{2m} - \alpha \cos(\beta x) - \gamma x. \tag{15}$$

This is often referred to as the cosine-Gordon equation, although it occurs more commonly with a sine term instead of the cosine term. Here, α is usually related to the d.c. Josephson current, which exists in the absence of an electric field in the Josephson tunnel junction [76]. The term β is related to the magnetic field, in which the Josephson current is oscillatory in the ratio of BA/Φ_0, where B is the magnetic field, A is the area of the junction, and Φ_0 ($=h/2e$) is the flux quantum (the factor of 2 arises as the tunneling particle is a Cooperon of paired electrons). The term $\gamma = eE$ is an applied electric field which gives a tilt to the periodice cosine potential. The cosine, or washboard, potential is commonly used in Josephson based qubits in quantum information.

To study this, we use the normalized parameters $m = 1.5625$, $\beta = 1.0$ and $\alpha = 4.0$. We use a numerical eigen solver routine to estimate the first three eigenvalues to be $E_0 = 0.816$, $E_1 = 3.25$, and $E_2 = 5.5$, all in normalized relative units. Instead of sampling the phase space coordinates directly, we instead sample the energy distribution and place the particles near the energy eigenvalue E_3 of the oscillator to generate the boson Fock state. For low values of γ, the system is essentially a bound well and should have eigenstates similar to a Fock state. As γ increases, the asymmetry should tilt and shift the eigen spectrum. We illustrate this with two values of γ, as shown in Figure 4. In panel (a), we initiate the Wigner function using the $n = 2$ Fock state, then let the particles evolve until a steady state Wigner function is reached. In panel (b), we raise the electric field to induce particles to diffuse into adjacent minima of the cosine potential, thus leading to quantum diffusion. In each case, as the tilt is increased, we see the Wigner function increasingly skewed from the excited harmonic oscillator we initialize the particles with.

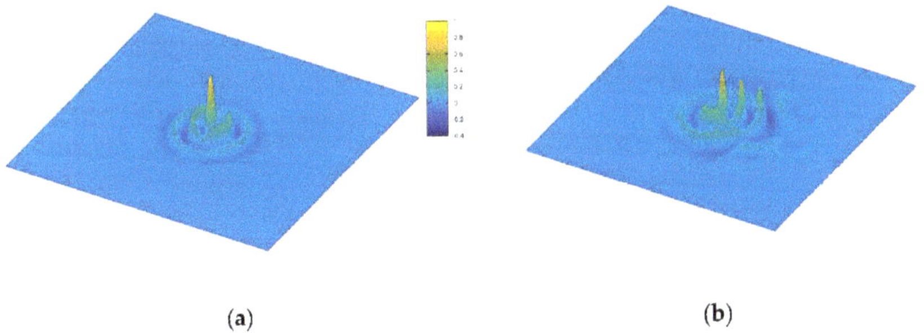

(a)   (b)

Figure 4. (a) Wigner function initialized near E_3 in the cosine potential, for the cosine-Gordon equation. Here, $\gamma = 0.2$. (b) Wigner function for $\gamma = 1.2$, illustrating quantum diffusion to adjacent cells of the cosine potential.

It was mentioned above that the tilted washboard potential is a model of the superconducting qubit Hamiltonian. It is straightforward to couple multiple qubits together with a pairwise interaction. Quantum circuits for quantum computing, such as the one recently unveiled by Google, involve only a small number of quantum particles, in their case 53 [77]. Generally, quantum simulation algorithms scale at best $O(\exp(N))$, where N is the number of particles. For the present algorithm, which is a full solver for the Wigner equation, the scaling is $O(N_p^3)$, where N_p is the number of pseudo-particles. The number of pseudo-particles is very challenging to estimate. The Wigner function is not a positive definite probability function, so one cannot use typical asymptotic convergence arguments to deduce the Monte Carlo convergence rate. Moreover, the number of particles per quantum particle in the present simulations is at least 10^5. In the worst case scenario, Google's quantum circuit would require

5.3×10^6 particles, assuming the qubit-qubit interaction does not necessitate far more particles for convergence. Such a simulation would be tractable on a small number of compute nodes, suggesting that simulating portions of the Google circuit using the above method is a plausible application to explore in the future.

5.4. Magnetic Single-Electron Control in an Interfering Double-Well Potential

Coherent single electron control is critically important for quantum information processing and advanced logic device operation principles based on the quantum character of the electron evolution on the nanometer scale [78]. An interesting and novel mechanism to coherently control single electron dynamics is provided by magnetic double-well potential structures [79]. Here, we investigate the effect of a uniform magnetic field on the electron state interference pattern manifesting in a double-well potential waveguide by means of full Wigner quantum transport simulations.

To be able to clearly investigate the electron quantum dynamics, we focus on coherent transport, i.e., no scattering processes are considered. In extension to previous work, here we target a non-focusing potential well setup, inspired by [80]. We consider the evolution of an initial electron state described by the Wigner function in a two-dimensional phase space $(r = x, y; p = k_x, k_y)$ in the presence of a uniform magnetic field B [81] and simulated by the Wigner EMC method using the signed-particle model via ViennaWD [82].

The governing evolution equation is obtained by introducing the magnetic component of the Lorentz force in analogy with the classical (Boltzmann) equation. Indeed, it is not an approximation but an exact quantum-coherent model obtained from the general magnetic Wigner theory for the case of a spatially-dependent but near stationary electric field $E(r)$ and a constant magnetic field B.

Figure 5 shows the principal details of the geometry of the simulated waveguide defined by infinite potentials along the left and right boundary as well as the averaged electron density distribution for symmetrically-sized potential wells and with, and without, a magnetic field. The boundaries in the vertical transport direction (top and bottom) are open. Green isolines at −0.15 eV indicate the potentials wells (Coulomb profile; peak potential at −0.35 eV). The initial state of the electron is the Wigner function corresponding to a minimum uncertainty wave packet with a standard deviation of $\sigma = 16$ nm . The central wave vector is $\left(k_{0x}, k_{0y}\right) = \left(0, 0.837 \text{ nm}^{-1}\right)$ and corresponds to an energy of 0.14 eV . The initial state is centered at $(x = 20$ nm $, y = 0$ nm$)$ and is injected at the bottom boundary, directed upwards towards the wells. Electrons are injected every femtosecond and do not interact with each other: The electron injections represent independent, identically distributed trials, similar to the Young-type double-slit experiments.

As has been previously shown and as expected, symmetric potential wells give rise to a symmetric electron density interference pattern. However, reducing a potential well by 50% bends the density pattern to the right and shows how the pattern can be manipulated by a potential well induced electric field (Figure 5a). A similar but stronger behavior is observed if the magnetic field is enabled: Figure 5b corresponds to the symmetric double-well potential case with an applied magnetic field ($B = -6$ T). The magnetic field shifts the density pattern in a more pronounced way than the asymmetric case. Both effects, the magnetic field and the asymmetric potentials, can be combined to work in tandem to further shift the pattern to the right.

However, despite the similarity of the electric and magnetic effects on the quantum electron density distribution, we observe that the electric and magnetic fields play a very different role in the transport dynamics. Figure 6 illustrates the Wigner function negativity maps corresponding to the setups shown in Figure 5. As previously mentioned, the Wigner function develops negative values in regions of quantum correlation effects. The maps $f_W^-(x, y)$ are created by integrating the negative values of the corresponding Wigner functions over the momentum coordinates. As Figure 5 clearly shows, the magnetic field destroys the coherence of the dynamics as the negativity is drastically reduced. This can be linked to the role of the two responsible electromagnetic terms in the transport equation: The action of E is independent from the particle momentum so that the particles in the ensemble are accelerated synchronously. To the contrary, the action of B explicitly depends on the momentum, which distorts the evolution.

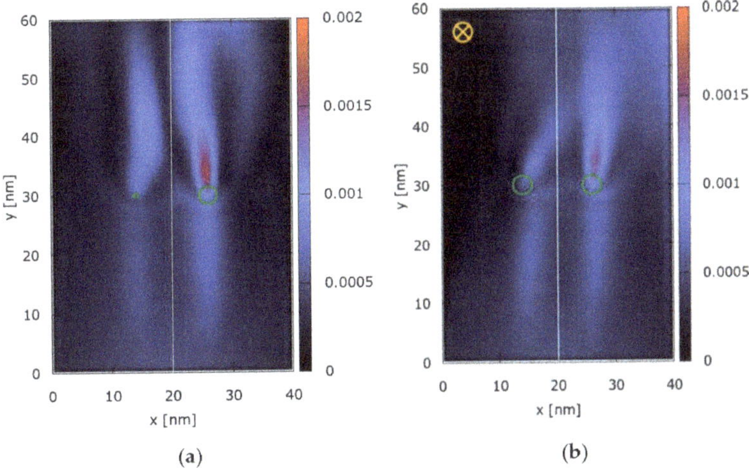

Figure 5. Averaged electron density (arbitrary units): (**a**) Asymetric potential wells and no magnetic field; (**b**) Symmetric potential wells and applied magnetic field. Green isolines indicate potential wells.

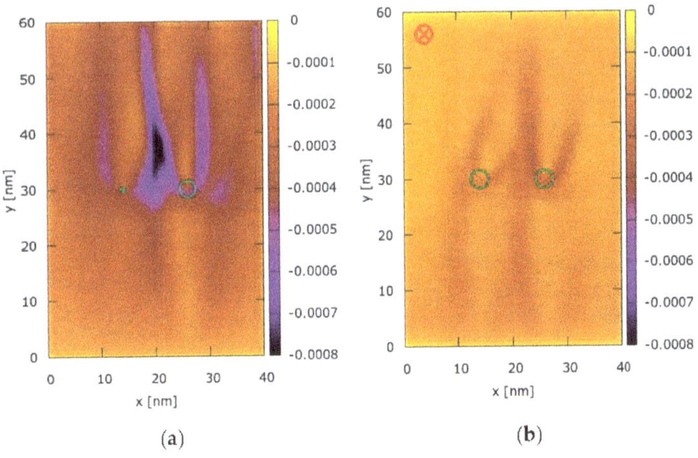

Figure 6. Wigner function negativity map: (**a**) Asymetric potential wells and no magnetic field; (**b**) Symmetric potential wells and applied magnetic field. Green isolines indicate potential wells.

6. Conclusions

Carrier transport in complex quantum systems, such as nanoscale semiconductor devices embedded within a complicated environment, requires a complex quantum description, which takes into account quantum-coherent phenomena, as well as dissipative processes of scattering, with both modified by the interaction with the environment. The use of the Wigner function is particularly useful in studies of these systems, as the Wigner function may explicitly illustrate the important quantum effects and entanglement that is a signature of quantum interactions. Particle approaches to Monte Carlo simulation of the quantum Wigner function provide an efficient approach to the study of such methods. In these methods, it becomes clear how to study each process and establish

Entropy **2020**, *22*, 1103

its importance in the behavior of the overall system. The Wigner function allows one to clearly identify the quantum effects, particularly the entanglement that arises between different parts of the quantum system.

Author Contributions: All authors contributed equally to the preparation of this manuscript. I.W. and D.K.F. performed the work for the second and third examples. J.W., M.N., M.B. and S.S. performed the work for the first and third examples. All authors have read and agreed to the published version of the manuscript.

Funding: The financial support by the Austrian Science Fund (FWF): P29406 and P33609, the Austrian Federal Ministry for Digital and Economic Affairs and the National Foundation for Research, Technology and Development is gratefully acknowledged.

Acknowledgments: The computational results presented here have been achieved in part using the Arizona State University Research Computing Center and the Vienna Scientific Cluster (VSC). D.K.F. and I.W. would also like to acknowledge discussions with S. M. Goodnick and D. Vasileska.

Conflicts of Interest: The authors have no conflict of interest.

References

1. Hisamoto, H.; Lee, H.-C.; Kedzierski, J.; Takeuchi, H.; Asano, K.; Kuo, C.; Andersen, E.; King, T.-J.; Bokor, J.; Hu, C.-M. FinFET-a self-aligned double-gate MOSFET scalable to 20 nm. *IEEE Trans. Electron.Dev.* **2000**, *47*, 2320–2325. [CrossRef]

2. Doyle, B.S.; Datta, S.; Doczy, M.; Hareland, S.; Jin, B.; Kavalieros, J.; Linton, T.; Murthy, A.; Rios, R.; Chau, R. High performance fully-depleted tri-gate CMOS transistors. *IEEE Electron. Dev. Lett.* **2003**, *24*, 263–265. [CrossRef]

3. Barker, J.R.; Ferry, D.K. On the physics and modeling of small semiconductor devices—II The very small device. *Sol. State Electron* **1980**, *23*, 531–544. [CrossRef]

4. Zurek, W.H. Decoherence, einselection, and the quantum origins of the classical. *Rev. Mod. Phys.* **2003**, *75*, 715–776. [CrossRef]

5. Prigogine, I. *From Being to Becoming: Time and Complexity in the Physical Sciences*; W. H. Freeman: San Francisco, CA, USA, 1980.

6. Ferry, D.K.; Barker, J.R. On the physics and modeling of small semiconductor devices—III Transient response in the finite collision-duration regime. *Sol. State Electron* **1980**, *23*, 545–549. [CrossRef]

7. Akis, R.; Ayubi-Moak, J.S.; Faralli, N.; Ferry, D.K.; Goodnick, S.M.; Saraniti, M. The upper limit of cutoff frequency in ultrashort gate-length InGaAs/InAlAs HEMTs: A new definition of effective gate length. *IEEE Electron. Dev. Lett.* **2008**, *29*, 306–308. [CrossRef]

8. Ferry, D.K. *Quantum Mechanics: An Introduction for Device Physicists and Electrical Engineers*, 2nd ed.; Sec. 7.4; Taylor and Francis: Bristol, UK, 2001.

9. Nakajima, S. On quantum theory of transport phenomena. *Prog. Theor. Phys.* **1958**, *20*, 948–959. [CrossRef]

10. Zwanzig, R. Ensemble method in the theory of irreversibility. *J. Chem. Phys.* **1960**, *33*, 1338–1341. [CrossRef]

11. Mori, H. Transport, collective motion, and Brownian motion. *Prog. Theor. Phys.* **1965**, *33*, 423–455. [CrossRef]

12. Barker, J.R. Quantum transport theory of high-field conduction in semiconductors. *J. Phys. C Sol. State Phys.* **1973**, *17*, 2663–2684. [CrossRef]

13. Ferry, D.K.; Akis, R.; Burke, A.M.; Knezevic, I.; Brunner, R.; Meisels, R.; Kuchar, F.; Bird, J.P. Open quantum dots: Physics of the non-Hermitian Hamiltonian. *Fortschr. Phys.* **2013**, *61*, 291–304. [CrossRef]

14. Okubo, Y.; Bird, J.P.; Ochiai, Y.; Ferry, D.K.; Ishibashi, K.; Aoyagi, Y.; Sugano, T. Magnetically induced suppression of phase breaking in ballistic mesoscopic billiards. *Phys. Rev. B* **1997**, *55*, 1368–1371. [CrossRef]

15. Ferry, D.K.; Akis, R.; Brunner, R. Probing the quantum-classical connection with open quantum dots. *Phys. Scr.* **2015**, *T165*, 014010. [CrossRef]

16. Brunner, R.; Akis, R.; Ferry, D.K.; Kuchar, F.; Meisels, R. Coupling-induced bipartite pointer states in arrays of electron billiards: Quantum Darwinism in action? *Phys. Rev. Lett.* **2008**, *101*, 024102. [CrossRef]

17. Tvisidis, Y. *Operation and Modeling of the MOS Transistor*, 2nd ed.; Oxford University Press: New York, NY, USA, 1998.

18. Ferry, D.K.; Akis, R.; Gilbert, M.J. Semiconductor device scaling: The role of ballistic transport. *J. Comput. Theor. Nanosci.* **2007**, *4*, 1149–1152. [CrossRef]

19. Ollivier, H.; Poulin, D.; Zurek, W.H. Objective properties from subjective quantum states: Environment as a witness. *Phys. Rev. Lett.* **2004**, *93*, 220401. [CrossRef] [PubMed]

20. Jacoboni, C.; Lugli, P. *The Monte Carlo Method for Semiconductor Device Simulation*; Springer: Vienna, Austria, 2005.

21. Schrödinger, E. Quantiseirung als Eigenwertproblem. *Ann. Phys.* **1926**, *79*, 361–376. [CrossRef]

22. Madelung, E. Quantentheorie in hydrodynamischer form. *Z. Phys.* **1927**, *40*, 322–327. [CrossRef]

23. Kennard, E.H. On the quantum mechanics of a system of particles. *Phys. Rev.* **1928**, *31*, 876–890. [CrossRef]

24. Shifren, L.; Akis, R.; Ferry, D.K. Correspondence between quantum and classical motion: Comparing Bohmian mechanics with a smoothed effective potential approach. *Phys. Lett. A* **2000**, *274*, 75–83. [CrossRef]

25. Brunner, R.; Meisels, R.; Kuchar, F.; Akis, R.; Ferry, D.K.; Bird, J.P. Classical and quantum mechanical simulations in open quantum dots. *J. Comput. Electron* **2007**, *6*, 93–96. [CrossRef]

26. Brunner, R.; Meisels, R.; Kuchar, F.; Akis, R.; Ferry, D.K.; Bird, J.P. Draining of the sea of chaos: Role of resonant transmission in an array of billiards. *Phys. Rev. Lett.* **2007**, *98*, 204101. [CrossRef] [PubMed]

27. Jacoboni, C.; Bertoni, A.; Bordone, P.; Brunetti, R. Wigner function formulation for quantum transport in semiconductors: Theory and Monte Carlo approach. *Math. Comp. Simulat.* **2001**, *55*, 67–78. [CrossRef]

28. Shifren, L.; Ferry, D.K. A Wigner function based ensemble Monte Carlo approach for accurate incorporation of quantum effects in device simulation. *J. Comp. Electron.* **2002**, *1*, 55–58. [CrossRef]

29. Nedjalkov, M.; Kosina, H.; Kosik, R.; Selberherr, S. A Wigner equation with quantum electron-phonon interaction. *Microelectr. Engr.* **2002**, *63*, 199–203. [CrossRef]

30. Zubarev, D. *Nonequilibrium Statistical Thermodynamics*; Springer: Berlin, Germany, 1974.

31. Ridley, B.K. Specific negative resistance in solilds. *Proc. Phys. Soc.* **1963**, *82*, 954–966. [CrossRef]

32. Gunn, J.B. Microwave oscillation of current in III-V semiconductors. *Sol. State Commun.* **1963**, *1*, 88–91. [CrossRef]

33. Poston, T.; Steward, I. *Catastrophe Theory and Its Applications*; Pitman: London, UK, 1978.

34. Wigner, E. On the quantum correction for thermodynamic equilibrium. *Phys. Rev.* **1932**, *40*, 749–759. [CrossRef]

35. Weinbub, J.; Ferry, D.K. Recent advances in Wigner function approaches. *Appl. Phys. Rev.* **2018**, *5*, 041104. [CrossRef]

36. Ferry, D.K.; Nedjalkov, M. *The Wigner Function in Science and Technology*; IOP Publishing: Bristol, UK, 2018.

37. Weyl, H. Quantenmechanik and Gruppentheorie. *Z. Phys.* **1927**, *46*, 1–46. [CrossRef]

38. Groenewold, H.J. On the principles of elementary quantum mechanics. *Physica* **1946**, *12*, 405–460. [CrossRef]

39. Moyal, J.E. Quantum mechanics as a statistical theory. *Math. Proc. Camb. Phil. Soc.* **1949**, *45*, 99–124. [CrossRef]

40. Diaz, N.C.; Prata, J.N. Admissable states in quantum phase space. *Ann. Phys.* **2004**, *313*, 110–146. [CrossRef]

41. Nedjalkov, M. Wigner transport in the presence of phonons: Particle models of the electron kinetics. In *Proceedings of the International School of Physics Enrico Fermi*; IOS Press: Amsterdam, The Netherlands, 2005; Volume 160, pp. 55–103. [CrossRef]

42. Nedjalkov, M.; Selberherr, S.; Ferry, D.K.; Vasileska, D.; Dollfus, P.; Querlioz, D.; Dimov, I.; Schwaha, P. Physical scales in the Wigner-Boltzmann equation. *Ann. Phys.* **2012**, *328*, 220–237. [CrossRef] [PubMed]

43. Bohm, D. A suggested interpretation of the quantum theory in terms of "hidden" variables I. *Phys. Rev.* **1952**, *85*, 166–179. [CrossRef]

44. Rossi, F.; Poli, P.; Jacoboni, C. Weighted Monte Carlo approach to electron transport in semiconductors. *Semicond. Sci. Technol.* **1992**, *7*, 1017–1036. [CrossRef]

45. Dimov, I.T. *Monte Carlo Methods for Applied Scientists*; World Scientific: Singapore, 1996.

46. Jacoboni, C.; Bordone, P. The Wigner function approach to non-equilibrium transport. *Repts. Prog. Phys.* **2004**, *67*, 1033–1072. [CrossRef]

47. Shifren, L.; Ferry, D.K. Particle Monte Carlo simulation of Wigner function tunneling. *Phys. Lett. A* **2001**, *285*, 217–221. [CrossRef]

48. Shifren, L.; Ferry, D.K. Wigner function quantum Monte Carlo. *Physica B* **2002**, *314*, 72–75. [CrossRef]

49. Shifren, L.; Ringhofer, C.A.; Ferry, D.K. A Wigner function-based quantum ensemble Monte Carlo study of a resonant tunneling diode. *IEEE Trans. Electron Dev.* **2003**, *50*, 763–773. [CrossRef]

50. Bordone, P.; Vasileska, D.; Ferry, D.K. Collision-duration time for optical-phonon emission in semiconductors. *Phys. Rev. B* **1996**, *53*, 3846–3855. [CrossRef] [PubMed]

51. Shifren, L.; Ferry, D.K. Inclusion of nonlocal scattering in quantum transport. *Phys. Lett. A* **2003**, *306*, 332–336. [CrossRef]
52. Ellinghaus, P.; Weinbub, J.; Nedjalkov, M.; Selberherr, S.; Dimov, I. Distributed-memory parallelization of the Wigner Monte Carlo method using spatial domain decomposition. *J. Comp. Electron.* **2015**, *14*, 151–162. [CrossRef]
53. Weinbub, J.; Ellinghaus, P.; Nedjalkov, M. Domain decomposition strategies for the two-dimensional Wigner Monte Carlo method. *J. Comp. Electron.* **2015**, *14*, 922–929. [CrossRef]
54. Querlioz, D.; Dollfus, P. *The Wigner Monte Carlo Method for Nanoelectronic Devices—A Particle Description for Quantum Transport and Decoherence*; Wiley-ISTE: London, UK, 2010.
55. Lousier, M.; Schenk, A.; Fichtner, W.; Klimeck, G. Atomistic simulation of nanowires in the $sp^3d^5s^*$ tight-binding formalism: From boundary conditions to strain calculations. *Phys. Rev. B* **2006**, *74*, 205323. [CrossRef]
56. Maassen, J.; Harb, M.; Michaud-Rioux, V.; Zhu, Y.; Guo, H. Quantum transport modeling from first principles. *Proc. IEEE* **2013**, *101*, 518–530. [CrossRef]
57. Martinez, A.; Kalna, K.; Sushko, P.V.; Shluger, A.L.; Barker, J.R.; Asenov, A. Impact of body-thickness dependent band structure on scaling of double-gate MOSFETs: A DFT/NEGF study. *IEEE Trans. Nanotechnol.* **2009**, *8*, 159–166. [CrossRef]
58. Zhou, W.X.; Chen, K.Q. Enhancement of thermoelectric performance by reducing phonon thermal condance in multiple core-shell nanowires. *Sci. Repts.* **2014**, *4*, 7150. [CrossRef] [PubMed]
59. Vasileska, D.; Prasad, C.; Wieder, H.H.; Ferry, D.K. Green's function approach for transport calculation in an $In_{0.53}Ga_{0.47}As/In_{0.52}Al_{0.48}As$ modulation-doped heterostructure. *J. Vac. Sci. Technol. B* **2003**, *21*, 1903–1907. [CrossRef]
60. Vasileska, D.; Eldridge, T.; Bordone, P.; Ferry, D.K. Quantum transport simulation of the DOS function, self-consistent fields, and the mobility in MOS inversion layers. *VLSI Des.* **1998**, *6*, 21–25. [CrossRef]
61. Meir, Y.; Wingreen, N.S. Landauer formula for the current through an interacting electron region. *Phys. Rev. Lett.* **1992**, *68*, 2512–2515. [CrossRef] [PubMed]
62. Keldysh, L.V. Diagram technique for nonequilibrium processes. *Sov. Phys. JETP* **1965**, *20*, 1018–1026.
63. Dyson, F.J. The radiation theories of Tomonaga, Schwinger, and Feynman. *Phys. Rev.* **1949**, *75*, 486–502. [CrossRef]
64. Bethe, H.; Salpeter, E. A relativistic equation for bound-state problems. *Phys. Rev.* **1951**, *84*, 1232–1242. [CrossRef]
65. Lugli, P.; Jacoboni, C.; Reggiani, L.; Kochevar, P. Monte Carlo algorithm for hot phonons in polar semiconductors. *Appl. Phys. Lett.* **1987**, *50*, 1251–1253. [CrossRef]
66. Schwaha, P.; Querlioz, D.; Dollfus, P.; St Martin, J.; Nedjalkov, M.; Selberherr, S. Decoherence effects in the Wigner function formalism. *J. Comp. Electron.* **2013**, *12*, 388–396. [CrossRef]
67. Xiao, D.; Liu, G.-B.; Feng, W.; Xu, X.; Yao, W. Coupled spin and valley physics in monolayers of MoS2 and other Group-VI dichalcogenides. *Phys. Rev. Lett.* **2012**, *108*, 196802. [CrossRef]
68. Lebègue, S.; Eriksson, O. Electronic structure of two-dimensional crystals from ab initio theory. *Phys. Rev. B* **2009**, *79*, 115409. [CrossRef]
69. Sinova, J.; Culcer, C.; Niu, Q.; Sinitsyn, N.A.; Jungwirth, T.; MacDonald, A.H. Universal intrinsic spin Hall effect. *Phys. Rev. Lett.* **2004**, *92*, 126603. [CrossRef]
70. Moca, C.P.; Marinescu, D.C. Fintie-size effects in a two-dimensional electron gas with Rashba spin-orbit interaction. *Phys. Rev. B* **2007**, *75*, 035325. [CrossRef]
71. Nikolic, B.K.; Souma, S.; Zarbo, L.B.; Sinova, J. Nonequilbrium spin Hall accumulation in ballistic semiconductor nanostructures. *Phys. Rev. Lett.* **2005**, *95*, 046601. [CrossRef]
72. Jungwirth, T.; Niu, Q.; MacDonald, A.H. Anomalous Hall effect in ferromagnetic semiconductors. *Phys. Rev. Lett.* **2002**, *88*, 207208. [CrossRef] [PubMed]
73. Xiao, D.; Yao, W.; Niu, Q. Valley contrasting physics in graphene: Magnetic moment and topological transport. *Phys. Rev. Lett.* **2007**, *99*, 236809. [CrossRef]
74. Ferry, D.K.; Welland, I. Relativistic Wigner functions in the transition metal di-chalcogenides. *J. Comp. Electron.* **2018**, *17*, 110–117. [CrossRef]
75. Welland, I.; Ferry, D.K. Wave-packet phase-space quantum Monte Carlo approach. *J. Comp. Electron.* in press.

76. Kevrekidis, P.G.; Daniela, I.; Caputo, J.-G.; Carretero-González, R. Planar and radial kinks in Klein-Gordon models: Existence, stability and dynamics. *Phys. Rev. E* **2018**, *98*, 052217. [CrossRef]

77. Arute, F.; Arya, K.; Babbush, R.; Bacon, D.; Bardin, J.C.; Barends, R.; Biswas, R.; Boixo, S.; Brandao, F.G.S.L.; Buell, D.A.; et al. Quantum supremacy using a programmable superconducting processor. *Nature* **2019**, *574*, 505–510. [CrossRef] [PubMed]

78. Bauerle, C.; Glattli, D.C.; Meunier, T.; Portier, F.; Roche, P.; Roulleau, P.; Takada, S.; Waintal, X. Coherent control of single electrons: A review of current progress. *Rpts. Prog. Phys.* **2018**, *81*, 056503. [CrossRef]

79. Ballicchia, M.; Nedjalkov, M.; Weinbub, J. Single electron control by a uniform magnetic field in a focusing double-well potential structure. In Proceedings of the 2020 IEEE 20th International Conference on Nanotechnology, Montreal, QC, Canada, 29–31 July 2020; pp. 73–76. [CrossRef]

80. Weinbub, J.; Ballicchia, M.; Nedjalkov, M. Electron interference in a double-dopant potential structure. *Phys. Stat. Sol. RRL* **2018**, *12*, 18000111. [CrossRef]

81. Nedjalkov, M.; Weinbub, J.; Ballicchia, M.; Selberherr, S.; Dimov, I.; Ferry, D.K. Wigner equation for general magnetic fields: The Weyl-Stratonovich transform. *Phys. Rev. B* **2019**, *99*, 014423. [CrossRef]

82. ViennaWD. Available online: http://www.iue.tuwien.ac.at/software/viennawd/ (accessed on 28 September 2020).

Article

Scattering in Terms of Bohmian Conditional Wave Functions for Scenarios with Non-Commuting Energy and Momentum Operators

Matteo Villani [1,†]**, Guillermo Albareda** [2,3,†]**, Carlos Destefani** [1,†]**, Xavier Cartoixà** [1,†] **and Xavier Oriols** [1,*,†]

1   Department of Electronic Engineering, Universitat Autònoma de Barcelona, Campus de la UAB, 08193 Bellaterra, Barcelona, Spain; matteo.villani@uab.es (M.V.); Carlos.Destefani@uab.es (C.D.); Xavier.Cartoixa@uab.es (X.C.)
2   Max Planck Institute for the Structure and Dynamics of Matter, Luruper Chaussee 149, 22761 Hamburg, Germany; guillermo.albareda@mpsd.mpg.de
3   Institute of Theoretical and Computational Chemistry, Universitat de Barcelona, Gran Via de les Corts Catalanes 585, 08007 Barcelona, Spain
*   Correspondence: xavier.oriols@uab.es
†   These authors contributed equally to this work.

Abstract: Without access to the full quantum state, modeling quantum transport in mesoscopic systems requires dealing with a limited number of degrees of freedom. In this work, we analyze the possibility of modeling the perturbation induced by non-simulated degrees of freedom on the simulated ones as a transition between single-particle pure states. First, we show that Bohmian conditional wave functions (BCWFs) allow for a rigorous discussion of the dynamics of electrons inside open quantum systems in terms of single-particle time-dependent pure states, either under Markovian or non-Markovian conditions. Second, we discuss the practical application of the method for modeling light–matter interaction phenomena in a resonant tunneling device, where a single photon interacts with a single electron. Third, we emphasize the importance of interpreting such a scattering mechanism as a transition between initial and final single-particle BCWF with well-defined central energies (rather than with well-defined central momenta).

Keywords: quantum dissipation; Bohmian mechanics; collision; conditional wave function; decoherence; open systems; many-body problem

Citation: Villani, M.; Albareda, G.; Destefani, C.; Cartoixà, X.; Oriols, X. Scattering in terms of Bohmian conditional wave functions for scenarios with non-commuting energy and momentum operators. *Entropy* **2021**, *23*, 408. https://doi.org/10.3390/e23040408

Academic Editor: Carlo Cafaro

Received: 4 February 2021
Accepted: 24 March 2021
Published: 30 March 2021

Publisher's Note: MDPI stays neutral with regard to jurisdictional claims in published maps and institutional affiliations.

1. Introduction

Due to the well-known many-body problem, electron transport in nanoscale devices must be modeled as an open quantum system [1]. The contacts, cables, atoms, electromagnetic radiation, etc. are commonly considered part of the environment. The effect of this environment on the dynamics of the simulated degrees of freedom, i.e., the electrons in the active region, can be recovered using some type of perturbative approximation. There are different formalisms in the literature to deal with such *environmental* perturbation (Green's functions [2–4], density matrix [5,6], Wigner distribution function [7–11], Kubo formalism [12], Pauli quantum Master equation [13,14], pure states [15,16], etc). In this work, we analyze the possibility of modeling the quantum nature of such simulated degrees of freedom with single-particle time-dependent pure states and their *environmental* perturbation as a transition between such single-particle time-dependent pure states.

In particular, we are interested in modeling the collision of an electron with a phonon or/and photon in an active region with tunneling barriers, i.e., in a scenario where the energy and momentum operators do not commute. The path to achieve this goal requires first the answer to the following question: *Is it possible to model an open system in terms of single-particle pure states?*. Once this conceptual question is answered, the next practical question that needs to be addressed is the following: *How do we select the single-particle pure states before and after the collision?* In this paper, we answer both questions. It will be shown that the alternative Bohmian formulation of quantum transport [17] provides a rigorous and versatile tool to describe collisions in open quantum systems in terms of single-particle time-dependent pure wave functions. This work is part of a long-term research project for the development of a general-purpose nanoelectronic device simulator, the so-called BITLLES simulator [18], using Bohmian trajectories.

The structure of the paper is the following. In Section 2, the answer to the first question about using single-particle pure states for open systems is provided from the Bohmian description of quantum phenomena. In Section 3, we provide an exact model for matter–light interaction in a closed system. Some simulation results are reported for different conditions of the total energy and a final discussion on the interaction between active region and environment to extend this description to an open system is provided. In Section 4, the practical implementation of the transition between pre-selected and post-selected states is discussed. This transition is performed for two different models: model A deals with energy conservation, and model B deals with momentum conservation. In Section 5 these two models, computed in a flat potential and in an arbitrary potential, are compared. Our conclusions are summarized in Section 6.

2. Is It Possible to Model an Open System in Terms of Single-Particle Pure States?

As we have stated, the active region of an electron device is, strictly speaking, an open quantum system interacting with the contacts, atoms in thermal motion, radiation, etc. As a consequence, in principle, one is not allowed to describe the electron in the active region in terms of pure states, but one has to rely on the use of the reduced density matrix.

Most approaches to open systems revolve around the reduced density matrix built by tracing out the degrees of freedom of the environment [1]. The ability to describe open systems with pure states can be partially justified when dealing with Markovian systems. In a pragmatical definition of Markovianity [19], the correlations between system and environment decay in a time scale that is much smaller than the observation (or relevant) time interval of the system. Thus, it can be assumed that every time we observe the system, it is defined by a pure state. For Markovian evolutions, the Lindblad master equation [20] for the reduced density matrix is a standard simulation tool. In addition, in Markovian scenarios where the off-diagonal terms of the reduced density matrix become irrelevant, a quantum master equation can be implemented, dealing with transitions between pure states [13,14].

In fact, it is possible to develop stochastic Schrödinger equations to unravel the reduced density matrix in terms of a pure-state solution for either Markovian or non-Markovian systems. The pure-state solution of stochastic Schrödinger equations can be interpreted as the state of the Markovian system while the environment is under (continuous) observation. However, such a physical interpretation cannot be given to the solutions of the stochastic Schrödinger equations for non-Markovian systems [21–30], where pure states can provide the correct one-time ensemble value but cannot be used to compute time correlations.

Therefore, for general non-Markovian quantum processes, when we are interested in a time-resolved description of the electron device performance, it is not possible to define the open system in terms of orthodox pure states. As described in [31] and explained below, a proper solution for treating electrons in non-Markovian open systems as single-particle pure states comes from the Bohmian formalism.

To explain how the Bohmian theory allows for a general description of a many-body quantum system in terms of wave functions, we consider a simplified scenario with only two degrees of freedom: one degree of freedom x belonging to the system plus one degree of freedom y belonging to the environment. Thus, the pure state in the position representation solution of the unitary Schrödinger equation is $\Psi(x,y,t)$. For each experiment, labelled by j, a Bohmian quantum state is defined by $\Psi(x,y,t)$ plus two well-defined trajectories,

$X^j[t]$ in the x-physical space and $Y^j[t]$ in the y-physical space. The role of the many-body wavefunction $\Psi(x,y,t)$ is guiding each trajectory $X^j[t]$ with a velocity that reads [17,32,33]

$$v^j_x[t] = \frac{dX^j[t]}{dt} = \frac{J_x(X^j[t],Y^j[t],t)}{|\Psi(X^j[t],Y^j[t],t)|^2} = \frac{1}{m^*}\frac{\partial S(x,y,t)}{\partial x}\bigg|_{x=X^j[t],y=Y^j[t]}, \tag{1}$$

where $J_x(x,y,t) = \hbar\,\mathrm{Im}\left(\Psi^*(x,y,t)\frac{\partial}{\partial x}\Psi(x,y,t)\right)/m^*$ is the current density with m^* the mass of the x-particle, and where $S(x,y,t)$ is the phase of the wave function written in polar form $\Psi(x,y,t) = |\Psi(x,y,t)|e^{iS(x,y,t)/\hbar}$. Analogous definitions are possible for the $Y^j[t]$ trajectory. By construction, the two positions $\{X^j[t],Y^j[t]\}$ in different $j = 1,...,W$ experiments are distributed (obeying quantum equilibrium [32,33]) at any time as

$$|\Psi(x,y,t)|^2 = \frac{1}{W}\sum_{j=1}^{W}\delta(x - X^j[t])\delta(y - Y^j[t]). \tag{2}$$

The identity in (2) requires $W \to \infty$. Numerically, we only require a large enough W to reproduce ensemble values given by the Born law in agreement with the orthodox theory. From a computational point of view, to ensure that (2) is satisfied at any time t, we only have to select the initial position $\{X^j[0],Y^j[0]\}$ at time $t = 0$ according to the distribution $|\Psi(x,y,0)|^2$.

The Bohmian theory opens the possibility to deal with a wave function of a subsystem through the concept of Bohmian conditional wave function (BCWF) [33,34]. The BCWF is defined for the x-degree of freedom during the j-th experiment as

$$\psi^j(x,t) \equiv \Psi(x,Y^j[t],t). \tag{3}$$

We emphasize that $\psi^j(x,t)$ provides a rigorous (Bohmian) definition of a single-particle wave function for an open system [32] that still includes the correlations with the other degrees of freedom y. Notice that the reason why the BCWF has a relevant role in Bohmian theory is because the trajectory $X^j[t]$ is equivalently guided by $\Psi(x,y,t)$ or by $\psi^j(x,t)$. In other words, the velocity $v^j_x[t]$ in (1) can be equivalently computed from the BCWF as

$$v^j_x[t] = \frac{dX^j[t]}{dt} = \frac{J^j_x(X^j[t],t)}{|\psi^j(X^j[t],t)|^2} = \frac{1}{m^*}\frac{\partial s^j(x,t)}{\partial x}\bigg|_{x=X^j[t]}, \tag{4}$$

where $|\psi^j(x,t)|^2 = |\Psi(x,Y^j[t],t)|^2$, $J^j_x(x,t) = \hbar\,\mathrm{Im}\left(\psi^{j*}(x,t)\frac{\partial}{\partial x}\psi^j(x,t)\right)/m^*$, and $s^j(x,t)$ is the angle of the BCWF in polar form $\psi^j(x,t) = |\psi^j(x,t)|e^{i\,s^j(x,t)/\hbar}$. Notice that we have not performed any approximation about the Markovianity of the quantum system in the definition of the BCWF. Thus, at the conceptual level, we conclude that any quantum open system can be analyzed in terms of single-particle pure states (i.e., BCWF) using the Bohmian formalism. This is a well-known result [31] and provides a definitive positive answer to the initial question: *Is it possible to model open system in terms of single-particle pure states?* Yes. Notice that the BCWF $\psi^j(x,t)$ will be a time-dependent function either because $\Psi(x,y,t)$ is a time-dependent function or because the trajectory $Y^j[t]$ is moving.

Let us discuss now a more realistic scenario with N electrons inside the active region with degrees of freedom $\{x_1,x_2,...,x_N\}$ that we want to simulate explicitly (for simplicity, each electron is assumed to be defined in a 1D space). There are, however, M environmental degrees of freedom $\{y_1,y_2,...,y_M\}$ that we do not want to simulate explicitly. The new many-body wave function of such a scenario is $\Psi(x_1,x_2,...,x_N,y_1,y_2,...,y_M)$, which is numerically inaccessible. We define $\bar{X}^j_i[t] = \{x^j_1[t],..,x^j_{i-1}[t],x^j_{i+1}[t],...,x^j_N[t]\}$ as the set of all Bohmian trajectories of the system except $x^j_i(t)$ for the i-particle in the j-experiment. Notice that we are dealing now with a superindex j indicating the experiment and subindex i indicating each particle in a given experiment. We also define $Y^j[t] = \{y_1^j[t], \ldots, y_M^j[t]\}$ as the set of all trajectories of the environment for the j-experiment. Then, the set of equations of motion of the resulting N single-electron BCWF $\psi^j(x_1, t) \equiv \Psi(x_1, \bar{X}_1^j[t], Y^j[t], t), \ldots, \psi^j(x_N, t) \equiv \Psi(x_N, \bar{X}_N^j[t], Y^j[t], t)$ inside the active region can be written as follows:

$$
i\hbar \frac{d\psi^j(x_1, t)}{dt} = \left[-\frac{\hbar^2}{2m} \nabla_{x_1}^2 + U_{eff}^j(x_1, t) \right] \psi^j(x_1, t)
$$

$$
\vdots
$$

$$
i\hbar \frac{d\psi^j(x_N, t)}{dt} = \left[-\frac{\hbar^2}{2m} \nabla_{x_N}^2 + U_{eff}^j(x_N, t) \right] \psi^j(x_N, t). \tag{5}
$$

The effective single-particle potential $U_{eff}^j(x_i, t) \equiv U_{eff}^j(x_i, \bar{X}_i^j[t], Y^j[t], t)$ is

$$
U_{eff}^j(x_i, t) = U^j(x_i, t) + V^j(x_i, t) + \mathcal{A}^j(x_i, t) + i\mathcal{B}^j(x_i, t), \tag{6}
$$

where $U^j(x_i, t)$ is an external potential acting only on the system degrees of freedom x_i, $V^j(x_i, t)$ is the Coulomb potential between x_i and the rest of particles at fixed positions $\bar{X}_i^j[t]$ and $Y^j[t]$, and $\mathcal{A}^j(x_i, t)$ and $\mathcal{B}^j(x_i, t)$ are potentials responsible for the remaining of quantum correlations between the degrees of freedom of the system and the environment [31]. A mandatory clarification is needed here. Are the set of BCWFs in (5) solving the many-body problem? No. If you want to use the coupled system of equations of motion of the N BCWF in (5) to describe a given experiment, first, you have to solve the Poisson (Gauss) equation to find $U^j(x_i, t)$ and $V^j(x_i, t)$ explicitly and, second, you have to know the exact solution of the many-body wave function $\Psi(x_1, x_2, \ldots, x_N, y_1, y_2, \ldots, y_M)$ to find $\mathcal{A}^j(x_i, t)$ and $\mathcal{B}^j(x_i, t)$ for all electrons [31]. The last step is numerically inaccessible. The merit of the system of equations in (5) is showing that such a type of solution to the many-body function exists and that we can look for educated guesses on the shape of $\mathcal{A}^j(x, t)$ and $\mathcal{B}^j(x, t)$ to provide reasonable approximations. Notice that a similar procedure is followed in Density Functional Theory: it shows a method to rewrite the many-body wave function in terms of single-particle wave functions, but the procedure requires knowledge of the exchange-correlation functional, which is only known once the many-body wave function is known. See further details and an explanation on $\mathcal{A}^j(x, t)$ and $\mathcal{B}^j(x, t)$ in [18,31–33,35,36].

To better appreciate the details of this simulation technique for electron devices, we notice that the total current $I^j(t)$ at time t for the j-experiment, after solving the set of BCWF from (5) with the appropriate approximations for $\mathcal{A}^j(x, t)$ and $\mathcal{B}^j(x, t)$, can be defined from the Bohmian trajectories with the help of a quantum version of the Ramo–Shockley–Pellegrini theorem [37] as follows:

$$
I^j(t) = \frac{e}{L} \sum_{i=1}^{n(t)} v_{x_i}^j(x_i^j[t], \bar{X}_i^j[t], Y^j[t]), \tag{7}
$$

where L is the distance between the two (metallic) contacts that define the active region, e is the electron charge (with sign), and $v_{x_i}^j(x_i^j[t], \bar{X}_i^j[t], Y^j[t])$ is the Bohmian velocity of the ith electron inside the active region in the j-experiment. Notice that the *observables* are computed from the trajectories (not from the BCWF) and that they are linked to a particular experiment j (which can be understood as a single configuration of the environment). The different possible values of $x_i^j[t], \bar{X}_i^j[t]$ and $Y^j[t]$ for the same (*preparation of the*) many-body wave function $\Psi(x_1, x_2, \ldots, x_N, y_1, y_2, \ldots, y_M)$ introduce the inherent quantum randomness in any experiment. As such, if one is interested in ensemble average values, one can repeat the calculation for all environment configurations $Y^j[t]$ and particle distributions $x_i^j[t]$ and $\bar{X}_i^j[t]$. Typically, in electronics, this ensemble average of the current $I^j(t)$ over many experiments

$j = 1, ..., \infty$ is interesting in evaluating DC values of the electrical current under ergodic assumptions. In the laboratory, however, a large time-average of the current $I^j(t)$ in a single j-experiment is usually performed. If one is interested in noise or time-correlations of the current at different times, $I^j(t_1)$ and $I^j(t_2)$, then the access to the individual experiment offered by the BCWF is very relevant.

Finally, we mention which are the computational advantages of this simulation framework. It is a microscopic description of the transport in the sense that it provides an individual description for each electron inside the active region. It provides a rigorous estimation (a part from the approximations for $\mathcal{A}^j(x_i, t)$ and $\mathcal{B}^j(x_i, t)$) to the quantum dynamics of electrons in the active region (open quantum system) for Markovian and non-Markovian systems. It is a versatile approach in the sense that it can simulate many different scenarios, from steady-state DC to transient and AC, including fluctuations of the current (noise). Notice that $I^j(t)$ in (7) includes the particle and displacement current, even at THz frequencies, when multi-time measurements are implicit. In this sense, we argue that the amount of information that this simulator framework can provide in the quantum regime is comparable to the predicting capabilities of the traditional Monte Carlo solution of the Boltzmann transport equation [38] in the semi-classical regime.

3. How Do We Select the Single-Particle Pure States Before and after the Collision?

To discuss how electron–photon scattering can be included in this simulation framework, we provide, first, an exact computation of the interaction between a single electron and a single photon in a closed system in terms of BCWF and Bohmian trajectories and, second, some indications on how such interaction can be modeled in an open system.

3.1. Exact Solution in a Closed System

The full quantum Hamiltonian $\hat{H} = \hat{H}_e + \hat{H}_\gamma + \hat{H}_I$ that describes light–matter interaction is given by the sum of the electron Hamiltonian $\hat{H}_e$, the electromagnetic field Hamiltonian $\hat{H}_\gamma$, and the electron–photon interaction Hamiltonian $\hat{H}_I$. In particular, for a single electron in a semiconductor, the position representation for $\hat{H}_e$ (assuming a 1D system for the electron with degree of freedom x) is given by

$$H_e = -\frac{\hbar^2}{2m^*}\frac{\partial^2}{\partial x^2} + V(x), \tag{8}$$

where $V(x)$ includes both the internal and external electrostatic potentials. See the blue electron wave packet and the scalar potential $V(x)$ for a double barrier region of length $2L_x$ in the horizontal x-axis of Figure 1a.

We consider that the electromagnetic field is described by a single mode with angular frequency ω inside a closed cavity of length $2L_M$. See the cyan mirrors in the horizontal x-axis of Figure 1a,b. A typical description of the electric field will be $E(x, t) \propto q \cos(kx - \omega t)$ with the wave vector $k = 2\pi/\lambda$ related to the angular frequency as $c = \omega/k$ with c being the speed of light. The variable q represents the instantaneous amplitude of the electromagnetic field along the polarization vector. Under the assumption $L_M \gg L_x$, meaning that the wave-length for the electromagnetic wave ($\approx$500 nm) is much larger than the active region ($\approx$20 nm), we can neglect the spatial dependence x of the electromagnetic field. Then, the Hamiltonian of the electromagnetic field in second quantization can be written as $\hat{H}_\gamma = \hbar\omega(1/2 + \hat{a}^\dagger\hat{a})$. The relationship between the now quantized amplitude of the electric field q and the creation $\hat{a}^\dagger$ and annihilation operators $\hat{a}$ is given by

$$\hat{a} = \sqrt{\frac{\omega}{2\hbar}}\left(q + \frac{\hbar}{\omega}\frac{\partial}{\partial q}\right) \quad , \quad \hat{a}^\dagger = \sqrt{\frac{\omega}{2\hbar}}\left(q - \frac{\hbar}{\omega}\frac{\partial}{\partial q}\right). \tag{9}$$

Then, the q-representation of $\hat{H}_\gamma$ is

$$H_\gamma = -\frac{\hbar^2}{2}\frac{\partial^2}{\partial q^2} + \frac{\omega^2}{2}q^2, \tag{10}$$

where the electromagnetic vacuum state with zero photons $|0\rangle$ solution of $\hat{H}_\gamma$ corresponds to the ground state of a harmonic oscillator $\psi_0(q) = \langle q|0\rangle$, while the state solution of $\hat{H}_\gamma$ with one photon corresponds to the first excited state of a harmonic oscillator $\psi_1(q) = \langle q|\hat{a}^\dagger|0\rangle$.

The interaction Hamiltonian in the dipole approximation can be written as $\hat{H}_I = -e\hat{x}\hat{E}$, where e is the (unsigned) electron charge and the electrical field operator is given by $\hat{E} = \epsilon(\hat{a} + \hat{a}^\dagger)$, with ϵ the strength of the electric field, or explicitly as

$$H_I = \alpha' xq, \tag{11}$$

where α', which depends on ϵ and other parameters of the cavity, controls the strength of the light–matter interaction. Finally, the wave function $\Psi(x, q, t)$ that describes the quantum nature of electrons and the electromagnetic field simultaneously in the q-representation is the solution to the following two-dimensional Schrödinger equation:

$$\begin{aligned}
i\hbar\frac{\partial\Psi(x,q,t)}{\partial t} = \ &-\ \frac{\hbar^2}{2m}\frac{\partial^2\Psi(x,q,t)}{\partial x^2} + V(x)\Psi(x,q,t) \\
&-\ \frac{\hbar^2}{2}\frac{\partial^2\Psi(x,q,t)}{\partial q^2} + \frac{\omega^2}{2}q^2\Psi(x,q,t) \\
&+\ \alpha' xq\Psi(x,q,t).
\end{aligned} \tag{12}$$

To simplify our discussion on emission and absorption of a photon by an electron, let us assume that only the zero photon state, $\psi_0(q) = \langle q|0\rangle = \langle q|\psi_0\rangle$, and the one photon state, $\psi_1(q) = \langle q|\hat{a}^\dagger|0\rangle = \langle q|\psi_1\rangle$, are relevant in our active region. Notice that we discuss the interaction of a single electron with a single photon in a closed system. Then, we can rewrite the wave function $\Psi(x, q, t)$ solution of (12) as

$$\Psi(x, q, t) = \psi_A(x, t)\psi_0(q) + \psi_B(x, t)\psi_1(q), \tag{13}$$

with

$$\psi_A(x, t) = \int \psi_0^*(q)\Psi(x, q, t)dq, \tag{14}$$

$$\psi_B(x, t) = \int \psi_1^*(q)\Psi(x, q, t)dq. \tag{15}$$

The equation of motion of $\psi_A(x, t)$ and $\psi_B(x, t)$ can be obtained by introducing the definition (13) into (12) and by using the orthogonality of $\psi_0(q)$ and $\psi_1(q)$ as follows:

$$i\hbar\frac{\partial\psi_A(x,t)}{\partial t} = -\frac{\hbar^2}{2m}\frac{\partial^2\psi_A(x,t)}{\partial x^2} + \left(V(x) + \frac{1}{2}\hbar\omega\right)\psi_A(x,t) + \alpha x\psi_B(x,t), \tag{16}$$

$$i\hbar\frac{\partial\psi_B(x,t)}{\partial t} = -\frac{\hbar^2}{2m}\frac{\partial^2\psi_B(x,t)}{\partial x^2} + \left(V(x) + \frac{3}{2}\hbar\omega\right)\psi_B(x,t) + \alpha x\psi_A(x,t), \tag{17}$$

where we defined $\alpha = \alpha'\int \psi_0(q)q\psi_1(q)dq$ and we assumed $\int \psi_0(q)q\psi_0(q)\,dq = \int \psi_1(q)q\psi_1(q)\,dq = 0$.

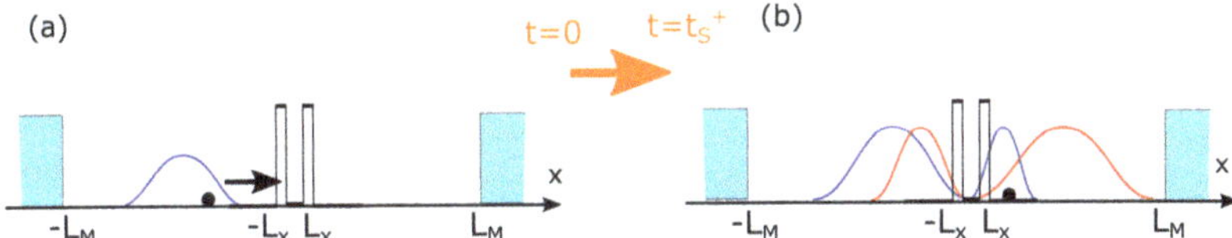

Figure 1. Schematic representation of the time evolution of the wave function for an electron impinging upon a double barrier region with electromagnetic radiation. In (**a**,**b**), we consider a cavity small enough so that the electromagnetic light does not radiate and so that no interaction with an environmental degree of freedom outside the active region is included. Only the information on the electron degree of freedom x and the internal degree of freedom of the light q (not plotted) are relevant. The Bohmian position of the electron $X[t]$ is indicated as a solid black circle. The $Q[t]$ trajectory of the electromagnetic field is not indicated. Notice that, in (**a**), the initial electron wave function is $\psi_A(x, t = 0) \neq 0$ (blue curve for the electron) and $\psi_B(x, t = 0) = 0$ while, in (**b**), we get $\psi_B(x, t = t_s^+) \neq 0$ (red curve) due to spontaneous emission.

We simulate now an initial electron impinging on a double barrier with a potential energy $V(x)$, as shown in Figure 2a. It corresponds to the conduction band of a typical resonant tunneling diode (RTD) with a 10 nm-well width, barrier thickness of 2 nm, and barrier height of 0.5 eV. In Figure 2b, the transmission coefficient of the double barrier is plotted, showing two resonant energies inside the well at $E_1 = 0.058$ eV and $E_2 = 0.23$ eV. The positive energies correspond to energy eigenstates impinging from the left and negative energies from the right side of the RTD device.

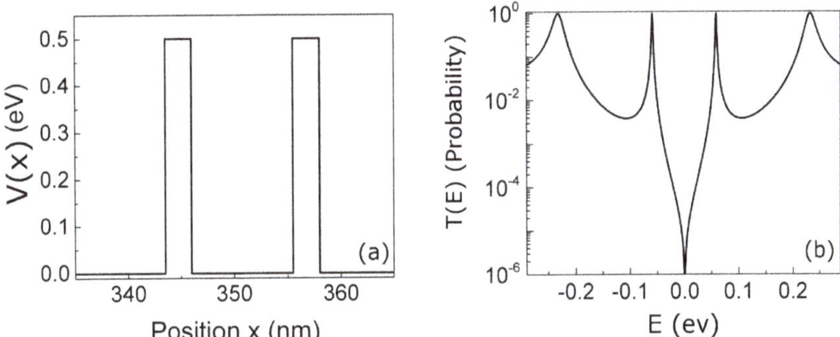

Figure 2. (**a**) Potential profile and (**b**) transmission coefficient T in function of injection energy E of a GaAs/AlGaAs resonant tunneling device (RTD) device with 10 nm well width. Positive energies means eigenstates injected from the left and negative energies eigenstates injected from the right.

At the initial time, we assume that there are no photons inside the active region. In other words, the (vacuum) electromagnetic field is given by an amplitude q with probability $|\psi_0(q)|^2$. Thus, we define $\psi_A(x, 0)$ as a Gaussian wave packet outside of the barrier region with a central energy equal to the second resonant level of the double barrier E_2 and a spatial dispersion of 30 nm, as seen in the blue wave packet in the x-axis of Figure 1a, and $\psi_B(x, 0) = 0$. Thus, the initial electron–photon wave function in Expression (13) is given only by $\Psi(x, q, t) = \psi_A(x, t)\psi_0(q)$. When solving (16) and (17) together, with $\alpha = 2.5 \cdot 10^7$ eV/m and $\omega = (E_2 - E_1)/\hbar$, we obtain that $\psi_B(x, t) \neq 0$ so that the global wave function in (13) becomes $\Psi(x, q, t) = \psi_A(x, t)\psi_0(q) + \psi_B(x, t)\psi_1(q)$. This process of spontaneous emission cannot be understood without the quantization of the electromagnetic field performed in (12).

Next, to compute how much probability inside the well can be assigned to $\psi_A(x,t)$ and $\psi_B(x,t)$, at each resonant level, we define

$$P_{A,1}(t) = \frac{1}{N}\int_0^{\frac{E_1+E_2}{2}} |c_A(E,t)|^2 dE \quad , \quad P_{A,2}(t) = \frac{1}{N}\int_{\frac{E_1+E_2}{2}}^{\infty} |c_A(E,t)|^2 dE, \tag{18}$$

with

$$c(E,t) = \int_{-L_x}^{L_x} \psi(x,t)\phi_E^*(x)dx, \tag{19}$$

The subindex A in $c(E,t)$ and $\psi(x,t)$ is assumed in (19). The functions $\phi_E(x)$ are the energy eigenstates of the electron Hamiltonian H_e in (8). Notice that we are only interested in the probability inside the barrier region with limits given by $x = \pm L_x$. Identical definitions can be provided for $P_{B,1}(t)$ and $P_{B,2}(t)$ with the normalization constant N, ensuring that $P_{A,1}(t) + P_{A,2}(t) + P_{B,1}(t) + P_{B,2}(t) = 1$.

In Figure 3, we plot $P_{A,1}(t)$, $P_{A,2}(t)$, $P_{B,1}(t)$, and $P_{B,2}(t)$, showing the typical Rabi oscillation. The initial value $P_{A,2}(0) \equiv 1$ in Figure 3 indicates an electron injected with a central energy equal to the second eigenvalue of the well without photons. The vertical dashed lines in Figure 3 indicate two times when the system passes from one electron in the first level and one photon ($P_{A,2} \approx 0$ and $P_{B,1} \approx 1$ in blue dashed line) to one electron in the second level and zero photons ($P_{A,2} \approx 1$ and $P_{B,1} \approx 0$ in red dashed line).

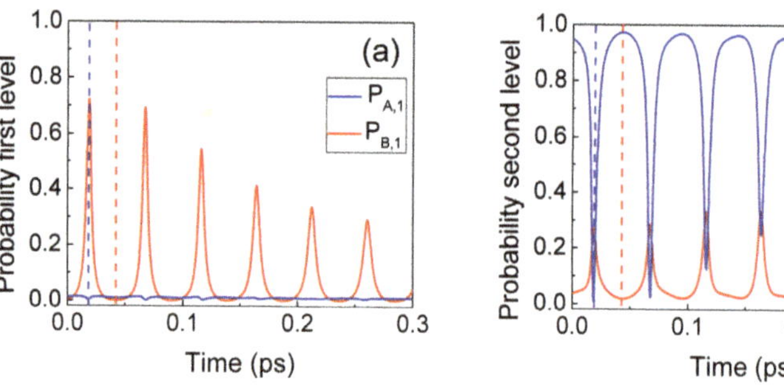

Figure 3. Evolution of the $P_{A,1}$, $P_{A,2}$, $P_{B,1}$, and $P_{B,2}$ for the first (**a**) and second (**b**) eigenstates of the quantum well described in Figure 2, when the Bohmian conditional wave function (BCWF) is injected in the second eigenstate of the quantum well.

From the whole wave function $\Psi(x,q,t) = \psi_A(x,t)\psi_0(q) + \psi_B(x,t)\psi_1(q)$, we can compute the probability presence in the x-space as follows:

$$P_e(x,t) = \int dq |\Psi(x,q,t)|^2 = |\psi_A(x,t)|^2 + |\psi_B(x,t)|^2. \tag{20}$$

In Figure 4a, we show the evolution of $P_e(x,t)$ computed from (20) as a function of time together with some selected trajectories $X^j[t]$. Such trajectories $X^j[t]$ are computed from the guiding total wave function $\Psi(x,q,t) = \psi_A(x,t)\psi_0(q) + \psi_B(x,t)\psi_1(q)$ together with the trajectories $Q^j[t]$ belonging to the electromagnetic degree of freedom q following the velocities defined in (1) for the same simulation presented before. The evolution of $P_e(x,t)$ inside the well shows qualitatively the alternate transition from one maximum (first eigenstate) to two maxima (second eigenstate). The Bohmian trajectories follow this evolution, since they alternatively move from one side to the center of the quantum well. The trajectories show a velocity close to zero when each eigenstate is well-defined and a large velocity during the transitions between the two eigenstates. All this dynamical information is in agreement with the physics of the Rabi oscillations depicted in Figure 3 where the electron emits a photon into a single-mode electromagnetic cavity and then reabsorbs it. As a technical detail, we mention that, as expected, Bohmian trajectories do not cross into the $x - q$ space (not plotted) but they cross in the subspace x of Figure 4a. In addition, one can expect some chaotic behavior in 2D systems [39,40] that is not present in the 1D system that is shown in the Figure 4a.

In Figure 4b, we plot the probability of the energy states $|c(E, t)|^2$ given by Equation (19) at the two times indicated by horizontal read and blue dashed lines in Figure 4a that correspond to the vertical dashed lines in Figure 3. The BCWF in Equation (19) has been defined as $\psi(x, t) = \Psi(x, Q^j[t], t) = \psi_A(x, t)\psi_0(Q^j[t]) + \psi_B(x, t)\psi_1(Q^j[t])$ for a selected trajectory $Q^j[t]$ of the j-experiment. Notice that such a definition of the BCWF corresponds to $\psi(x, t) \approx \psi_B(x, t)$ for the blue wave packet while the red wave packet corresponds to $\psi(x, t) \approx \psi_A(x, t)$ because of the values of $P_{A,2}$ and $P_{B,1}$ indicated by vertical dashed lines in Figure 3.

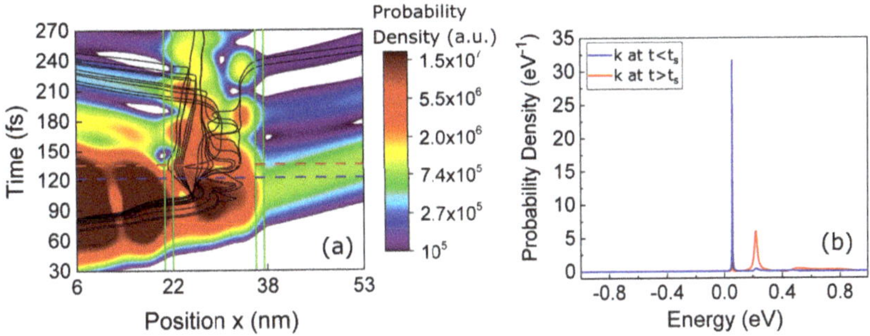

Figure 4. (**a**) Evolution of $P_e(x, t)$ for the electron interacting with the RTD device described in Figure 2, while emitting and absorbing electromagnetic radiation. The solid black lines show Bohmian trajectories $X^j[t]$ for a selected set of experiments. The green vertical lines indicate the position of the potential barriers. (**b**) Probability distribution of the Hamiltonian eigenstates for the BCWF given by $\psi(x, t) = \Psi(x, Q^j[t], t) = \psi_A(x, t)\psi_0(Q^j[t]) + \psi_B(x, t)\psi_1(Q^j[t])$ for a selected trajectory $Q^j[t]$ at two different times indicated by the horizontal dashed lines in (**a**). We define the scattering time t_s as the time of the blue horizontal dashed line.

As expected, the fact that the conservation of the total energy has to be satisfied from (12) has important consequences on the type of electron–photon interaction allowed. We now repeat the simulation when the electron (with no photon) is injected with a central energy corresponding to the first resonant level. No electron transition (or spontaneous emission) takes place, giving $\psi_B(x, t) \approx 0$ because the initial energy $E_1 + \hbar\omega/2$ cannot be converted into a much higher final energy $E_2 + 3\hbar\omega/2$. The result is shown in Figure 5.

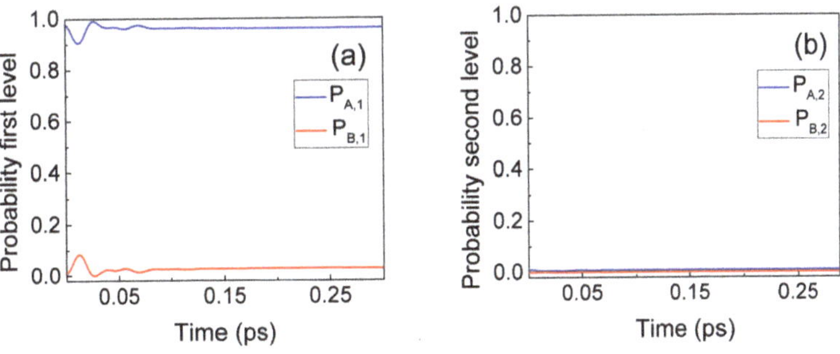

Figure 5. Evolution of $P_{A,1}$, $P_{A,2}$, $P_{B,1}$, and $P_{B,2}$ for the first (**a**) and second (**b**) eigenstates of the quantum well when the initial electron is injected in the first resonant level of the quantum well. Because of the conservation of energy, no matter–light interaction is possible.

We now repeat the same simulation done in Figure 3, where the initial electron had a mean energy equal to the second eigenvalue of the well, E_2, but considering a new photon energy $\hbar\omega = 0.26$ eV much larger than $E_2 - E_1 = 0.172$ eV. In this case, no light–matter interaction takes place since it would imply a violation of the conservation of whole energy. The initial energy $E_2 + \hbar\omega/2$ does not coincide with a possible final energy $E_1 + 3\hbar\omega/2$. This simulation is shown in Figure 6.

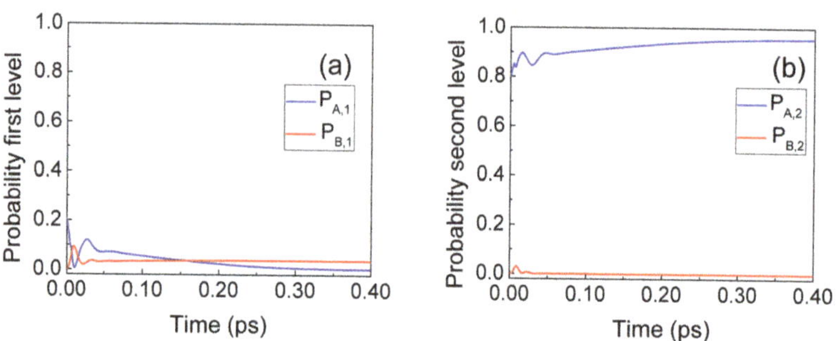

Figure 6. Evolution of $P_{A,1}$, $P_{A,2}$, $P_{B,1}$, and $P_{B,2}$ for the first (**a**) and second (**b**) eigenstates of the quantum well when the BCWF is injected in the second eigenstate of the quantum well and $\hbar\omega = 0.26$ eV. Because of the conservation of energy, no matter–light interaction is possible.

3.2. Approximate Solution with BCWF for an Open System

In the previous subsection, we discussed the interaction of a single electron with a single photon in a closed system. Here, we discuss how such an interaction can be generalized to include the possibility to detect a photon at a position y, far from the active region.

The proper simulation of such a scenario as a closed system is far from the scope of the present paper. Apart from considering the detector outside of the active region as a new electron with degree of freedom y, the transition of the electromagnetic energy from the active region to the environment will require an electromagnetic field with an arbitrary shape different from the one considered in the previous section. A Fourier transform of such an arbitrary electromagnetic field will imply dealing with several components $E(x, t) \propto q \, cos(kx - \omega t)$ at different frequencies. In any case, without an explicit solution of such problem, only from the conservation of energy, we can anticipate what will be the expected behavior of the whole system.

The process of spontaneous emission of a photon inside the active region and its posterior detection far from the active region can be anticipated as follows:

- At the initial time, $t = 0$, we consider an electron in the active region, with degree of freedom x with a central energy E_2 linked to zero photons wave function $\psi_0(q)$ plus another electron far from the active region with degree of freedom y and energy E_{ext} linked to zero photons $\psi_0(q)$. At this initial time, the total energy involved in such a scenario is $E_2 + \hbar\omega/2$ in the active region plus the energy $E_{ext} + \hbar\omega/2$ outside.

- At the intermediate time, we consider that a spontaneous emission of a photon happens inside the active region. As seen in Figure 3, such an internal process ensures energy conservation. Therefore, the new photon inside the active region implies a change in energy there, $E_2 + \hbar\omega/2 \rightarrow E_1 + 3\hbar\omega/2$, while the energy outside of the active region remains the same as before, $E_{ext} + \hbar\omega/2$. The total energy is the same as the initial one.

- At the final time t, we detect a photon at position y, far from the active region. Thus, the electron at y is now linked to one photon wave function $\psi_1(q)$, which implies an increment in the energy of $\hbar\omega$ far from the active region, $E_{ext} + \hbar\omega/2 \rightarrow E_{ext} + 3\hbar\omega/2$. The conservation of the total energy implies that the same amount of energy is eliminated in the active region when the photon leaves, $E_1 + 3\hbar\omega/2 \rightarrow E_1 + \hbar\omega/2$. The electron in the active region will have a new energy E_1 linked to the zero photons wave function $\psi_0(q)$. As we have seen in Figure 5, under such new energy conditions in the active region, such an electron will no longer be able to generate spontaneous emissions inside the RTD. Thus, the Rabi oscillations seen in Figure 3 for a closed system will not be present when we assume that the photon leaves the cavity.

In summary, we conclude that the spontaneous emission in the active region can be modeled by an initial BCWF $\psi(x,0)$ with central energy E_2 that changes to a final BCWF $\psi(x,t)$ with energy E_1. Such a process will be allowed as far as the photon energy coincides with $E_2 - E_1$. Identically, absorption in the active region can be modeled by an initial BCWF $\psi(x,0)$ with central energy E_1 that changes to a final BCWF $\psi(x,t)$ with energy E_2 with the photon energy given by $E_2 - E_1$. Notice that the conservation of energy enables the photon absorption to be accompanied, for example, by a subsequent process of spontaneous emission that returns the photon energy to the environment outside of the active region.

4. Implementation of the Transition from Pre- to Post-Selected BCWF

In this section, we describe practical issues on how such types of transitions between initial and final states can be implemented in a transport simulator for real electron devices based on Bohmian mechanics.

To implement the transition from pre- to post-selected BCWF, a definition of the initial $|i\rangle$ and the final $|f\rangle$ states is needed. Although the contacts do not allow us to perfectly *prepare* the wave description of electron, we can have some reasonable arguments to anticipate some of its properties. One option could be to deal with Hamiltonian eigenstates, which extend to infinite in both sides (left and right) of the device. Although these infinitely extended states are useful tools to model (steady-state) DC transport properties of quantum devices, they are less useful in describing other device performances, for example, the fluctuations of the electrical current due to the partition noise in a tunneling barrier. The initial electron, after impinging with the barrier, is either located to the left (reflection) or to the right (transmission) of a barrier but not on both sides of it. Such randomness (transmission or reflection) translates into current fluctuations. To model such fluctuations, a localized wave function seems appropriate to model electrons. However, the wave function cannot have a very narrow localization in position since the Heisenberg uncertainty principle would lead to extremely large momentum and energy uncertainties (larger than thermal energies). Thus, a definition of an electron, deep inside the contact, as a Gaussian wave packet with well-defined central position and central energy seems reasonable. We add that such a limited spatial extension of the electron wave function can be related to the coherence length of the sample.

In classical mechanics, an electron with a well-defined energy is compatible with an electron with a well-defined momentum. However, this is not the case for quantum electrons. As a general rule, two properties can be simultaneously well-defined if their operators commute. In our case, the energy (linked to the Hamiltonian operator $\hat{H}$) and the momentum (linked to the momentum operator $\hat{p}$) can be simultaneously defined when $[\hat{H},\hat{p}] = 0$. In the position representation, knowing that the Hamiltonian operator is the sum of the kinetic energy operator $(\hat{p})^2/2m$, which obviously commutes with $\hat{p}$, plus the potential energy operator $\hat{V}$, momentum and energy are well-defined properties when

$$\left[H, -i\hbar\frac{\partial}{\partial x}\right] = \left[V(x), -i\hbar\frac{\partial}{\partial x}\right] = i\hbar\frac{\partial V(x)}{\partial x} = 0. \tag{21}$$

Thus, only when dealing with flat potentials we can assume that a wave packet with a reasonable well-defined energy has also a reasonable well-defined momentum. This restriction seems relevant to transport models developed in phase-space (the Wigner distribution function), where information on only momenta and positions are available.

In the next two subsections, we discuss the implementation of the transition from a pre to a post-selected BCWF when using well-defined energies (model A) or momenta (model B). In Section 5, we compare the numerical results of these two different implementations.

4.1. Model A: Change in the Central Energy

We consider an electron defined by a single-particle BCWF that, at time t_s, undergoes a scattering event. We define $t_s^- = t_s - \Delta t_s$ as the time just before and $t_s^+ = t_s + \Delta t_s$ as the time just after the scattering event. For simplicity, we consider $\Delta t_s \to 0$, but we have seen in Section 3 that such a transition between initial and final BCWFs takes a finite time because, from a conceptual point of view, it has to guarantee the continuity of the BCWF in space and time. The initial and final BCWFs are $\psi(x, t_s^-)$ and $\psi(x, t_s^+)$, which satisfy $\langle E(t_s^+)\rangle = \langle E(t_s^-)\rangle + E_\gamma$, with E_γ the energy of a photon. Within the energy representation, the wave packet can be decomposed into a superposition of Hamiltonian eigenstates $\phi_E(x)$ of the electron $\hat{H}_e$ in (8) as

$$\psi(x, t_{s-}) = \int dE\, a(E, t_s^-)\, \phi_E(x),\tag{22}$$

with $a(E, t) = \int dx\, \psi(x, t)\, \phi_E^*(x)$. The central energy $\langle E(t_s^-)\rangle$ is

$$\langle E(t_s^-)\rangle = \int dE\, E\, |a(E, t_s^-)|^2,\tag{23}$$

which can be increased to obtain the new central energy at t_s^+ as

$$\begin{aligned}\langle E(t_s^+)\rangle &= \langle E(t_s^-)\rangle + E_\gamma = \int dE\,(E + E_\gamma)\,|a(E, t_s^-)|^2 = \int dE'\, E'\, |a(E' - E_\gamma, t_s^-)|^2 \\ &= \int dE'\, E'\, |a'(E', t_s^+)|^2,\end{aligned}\tag{24}$$

where we have defined $a'(E, t_s^+) = a(E - E_\gamma, t_s^-)$. Thus, the new wavepacket after the collision is

$$\psi(x, t_{s+}) = \int dE\, a'(E, t_s^+)\, \phi_E(x) = \int dE\, a(E' - E_\gamma, t_s^-)\, \phi_E(x).\tag{25}$$

This transition corresponds to the absorption of energy by the electron. Emission can be identically modeled by using $\langle E(t_s^+)\rangle = \langle E(t_s^-)\rangle - E_\gamma$. If required, the *technical* discontinuity between $\psi(x, t_{s-})$ and $\psi(x, t_{s+})$ can be solved by assuming that the change in energy is produced in a finite time interval $\Delta t_s = N_{t_s}\Delta t$, with Δt being the time step of the simulation. Then, at each time step of the simulation, the change in the wave packet central energy is E_γ/N_{t_s}. A *continuous* change in both energy and wave packet will be obtained as far as $\Delta t \to 0$. This continuous evolution of the BCWF can be represented as a Schrödinger-like equation, as explained in [31].

4.2. Model B: Change in Central Momentum

In Reference [34], we explain how a change in momentum p_γ in a wave packet in free space can be performed with a unitary Schrödinger equation. That algorithm can be understood as a pre- and a post-selection of the initial BCWF, $\psi(x, t_s^-)$, and of the final BCWF, $\psi(x, t_s^+)$, respectively. At time t_s^-, the BCWF can be written as a supersposition of momentum eigenstates $\phi_p(x)$ (which are a basis of the electron in the x space) as

$$\psi(x, t_{s-}) = \int dp\, b(p, t_{s-})\, \phi_p(x),\tag{26}$$

with $b(p, t_{s-}) = \int dx \psi(x, t_{s-})\, \phi_p^*(x)$. The central momentum $\langle p(t_s^-)\rangle$ is

$$\langle p(t_s^-)\rangle = \int dp\, p\, |b(p, t_{s-})|^2,\tag{27}$$

which can be increased to get the new central momentum $\langle p(t_s^+)\rangle = \langle p(t_s^-)\rangle + p_\gamma$ at t_s^+ as

$$
\begin{aligned}
\langle p(t_s^+)\rangle &= \langle p(t_s^-)\rangle + p_\gamma = \int dp\,(p + p_\gamma)\,|b(p,t_{s-})|^2 = \int dp'\,p'\,|b(p' - p_\gamma, t_{s-})|^2 \\
&= \int dp'\,p'\,|b(p',t_{s+})|^2,
\end{aligned}
\tag{28}
$$

where we have defined $b(p,t_{s+}) = b(p - p_\gamma, t_{s-})$. In this particular scenario, we know the explicit shape of the momentum eigenstates, $\phi_p(x) = 1/\sqrt{2\pi}\exp(ipx/\hbar)$, so that

$$
\begin{aligned}
\psi(x,t_s^+) &= \int dp\, b(p,t_{s+})\,\phi_p(x) = \int dp\, b(p - p_\gamma, t_{s-})\,\phi_p(x) \\
&= \int dp \int dx'\,\psi(x,t_{s-})\,\phi^*_{p-p_\gamma}(x')\phi_p(x) \\
&= \int dp \int dx\,\psi(x,t_{s-})\,\frac{1}{2\pi}e^{ip(x'-x)/\hbar}e^{ip_\gamma x'/\hbar} = e^{ip_\gamma x/\hbar}\psi(x,t_{s-}).
\end{aligned}
\tag{29}
$$

With the condition $\psi(x,t_s^+) = e^{ip_\gamma x/\hbar}\psi(x,t_{s-})$, it can be easily found the unitary equation satisfied by the BCWF. If we define $\psi'(x,t)$ as the wave function solution of the following Schrödinger equation, $i\hbar\frac{\partial\psi'(x,t)}{\partial t} = \frac{1}{2m^*}\left(-i\hbar\frac{\partial}{\partial x}\right)^2\psi'(x,t) + V(x)\psi'(x,t)$, with initial condition at $t = t_s$ given by $\psi'(x,t_s) = \psi(x,t_s^+)$, then the solution $\psi'(x,t)$ for $t > t_s$ is identical to the following Schrödinger equation, $i\hbar\frac{\partial\psi(x,t)}{\partial t} = \frac{1}{2m^*}\left(-i\hbar\frac{\partial}{\partial x} + p_\gamma\right)^2\psi(x,t) + V(x)\psi(x,t)$, for the original $\psi(x,t)$ and with its original initial condition for $t > t_s$. Finally, a single equation for $\psi(x,t)$ valid for all times is simply

$$
i\hbar\frac{\partial\psi(x,t)}{\partial t} = \frac{1}{2m^*}\left(-i\hbar\frac{\partial}{\partial x} + p_\gamma\Theta_{t_s}\right)^2\psi(x,t) + V(x)\psi(x,t),
\tag{30}
$$

where Θ_{t_s} is a Heaviside function equal to 1 for $t > t_s$ and zero otherwise. Thus, a description of the evolution of the wave function $\psi(x,t)$ during the collision process can be made from a unitary Schrödringer equation, where the momentum operator $-i\hbar\frac{\partial}{\partial x}$ is changed for the new momentum operator $-i\hbar\frac{\partial}{\partial x} + p_\gamma\Theta_{t_s}$, as indicated in [34]. Notice that the probability presence of the scattered wave packet satisfies $|\psi(x,t_{s+})|^2 = |\psi(x,t_{s-})|^2$ because only a global phase $e^{ip_\gamma x/\hbar}$ is added.

It is quite easy to see from (4) that the Bohmian velocity of the electron after the collision computed from $\psi(x,t_s^+)$ is just the old velocity computed from $\psi(x,t_s^-)$ plus p_γ/m^*,

$$
v_x^j[t_s^+] = \frac{1}{m^*}\frac{\partial s(x,t_s^+)}{\partial x}\Big|_{x=X^j[t]} = \frac{1}{m^*}\frac{\partial s(x,t_s^-)}{\partial x}\Big|_{x=X^j[t]} + p_\gamma/m^*.
\tag{31}
$$

The collision increases the velocity of the electron by the same amount that we add in (30). Unfortunately, as discussed at the beginning of the section, a global mechanism of scattering valid for scenarios with potential barriers requires dealing with a change in the energy as presented in model A (not with change of the momentum as presented in this model B).

5. Numerical Results

We present now the numerical results of our two models for the transition between initial and final single-particle BCWF, as explained in the previous section. We first study electron–photon collisions in free space, when energy and momentum operators commute, and then electron–photon collisions in a scenario with a double barrier potential profile, when energy and momentum operators do not commute. This last case will be compared with numerical results of the exact model presented in Section 3 and used to verify the physical soundness of the two models.

5.1. Collisions in Flat Potentials

In this section, we study the interaction of an electron and a photon in free space. The electron evolves in a flat potential. We consider the absorption of a photon by an electron. In flat potential, the momentum and energy conservation is ensured during the collision. Thus, since the momentum of the photon is negligible, in this section, we assume that the electron interacts with a phonon and a photon. The phonon will not be needed in Section 5.2. We consider that the final BCWF will be modeled by a final electron (post-selected state) with an energy increase of $\hbar\omega$ ($E_\gamma > 0$) plus the corresponding increase of momentum (provided by the phonon) with respect to the initial electron energy (pre-selected state).

In Figure 7, we show the simulation of the electron–photon collision in a flat potential. The collision is modeled by exchanging the energy $E_\gamma = 0.1$ eV in Figure 7a,b and by exchanging the momentum $p_\gamma = \sqrt{2E_\gamma/m^*}$ in Figure 7c,d. As expected, in this scenario, both models give identical results. After the scattering event, the Gaussian wave function evolves with a higher velocity, as indicated in (31). We notice that the wave function suffers a continuous evolution during the collision because it is a solution of the Schrödinger-like Equation (30). Analogous results (not shown) are obtained for emission. The main conclusion of this subsection is that model A and model B are, as expected, numerically equivalent in the case of a flat potential.

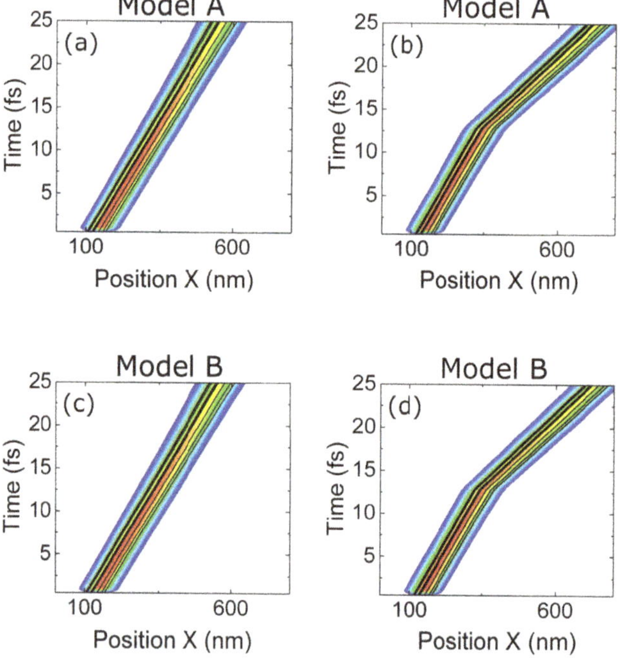

Figure 7. The evolution of the BCWF $\psi^j(x,t)$, undergoing photon absorption with $E_\gamma = 0.1$ eV, shown as function of position and time. The wavefunctions are simulated (**a**) without collision and (**b**) with collision using model A, and (**c**) without scattering and (**d**) scattered using model B. The trajectories $X^j[t]$ guided by the BCWF $\psi^j(x,t)$, where $j = 1,\ldots,10$, are some representative experiments and are shown in black. In a flat potential, the results of models A and B are identical.

5.2. Collisions in Arbitrary Potentials

As in Section 5.1, we study the absorption of a photon using an electron modeled by a final electron (post-selected state) with an energy increase of $\hbar\omega$ ($E_\gamma > 0$) with respect to the initial electron energy (pre-selected state). Now, we use a double barrier potential $V(x)$ identical to the one mentioned in Section 3.1, with the same two resonant energies $E_1 = 0.058\,\text{eV}$ and $E_2 = 0.23\,\text{eV}$.

In Figure 8, the evolution of $\psi^j(x,t)$ and the trajectories $X^j[t]$ are shown when the electron absorbs a photon while impinging on the potential barrier of the RTD. The position of the barriers is shown by the green vertical lines. The energy of the photon is equal to the difference of the resonant energies in the quantum well, $E_\gamma = E_2 - E_1$, and the BCWF is injected with a central energy equal to the first resonant energy $E = E_1$. A transition from E_1 to E_2 is expected during the collision $\psi_A(x,t_s^-) \to \psi_B(x,t_s^+)$.

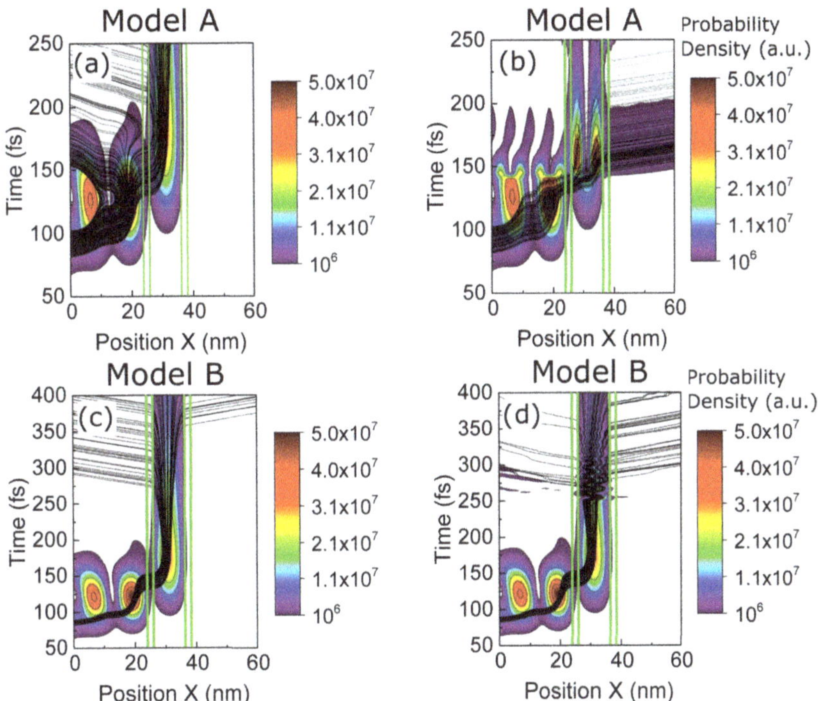

Figure 8. Gaussian wavefunctions interacting with a double barrier potential profile with and without scattering with a photon: (**a**) a wavepacket and some selected trajectories with unitary evolution (without scattering). (**b**) The same wave packet and the same selected trajectories when scattering with energy $E_\gamma = 0.186\,\text{eV}$ using model A occurs. (**c,d**) are identical to (**a,b**) when model B is used. In all figures, the Gaussian wave packet is injected from the left at energy $E = E_1 = 0.058\,\text{eV}$. The trajectories $X^j[t]$ guided by the BCWF $\psi^j(x,t)$ are plotted in black. The set of trajectories in plot (**a**) is different from the one in plot (**c**), with the goal of selecting those trajectories that interact most in the quantum well in each case. The trajectories in plot (**b**) are the same as in plot (**a**), and the trajectories in plot (**d**) are the same as in plot (**c**). The energy of the photon is equal to the distance between the two first energy levels, $E_\gamma = E_2 - E_1$.

In Figure 8a, we plot the time evolution of the electron interacting with the barrier but without photon collision. In Figure 8b, an electron–photon collision is produced at $t_s = 150$ fs using model A. The wavepacket undergoes a shift in energy probability distribution of the Hamiltonian eigenstates $\phi_E(x)$ towards higher values. As expected, the evolution of $\psi(x, t)$ is a transition from the first eigenstate of the well (with one peak of probability in the middle of the well) to the second one (with two probability peaks). The same trajectories $X^j[t]$ that were first reflected by the barrier in Figure 8a are now transmitted through the well in Figure 8b because the second resonant level has a wider transmission probability, as shown in Figure 2b. The results in Figure 8b have a reasonable agreement with the results in Figure 4a at times equivalent to the blue and red horizontal lines of Figure 4a. Clearly, we also notice that the simulated result in Figure 4a belongs to a simulation with the active region as a closed system, where the photon energy does not disappear, and the electron is continuously emitting and absorbing such photon energy, as explained in Section 3.1. On the contrary, Figure 8b corresponds to a simulation of the active region as an open system, where the photon energy appears/disappears at/from the active region only once, as explained in Section 3.2.

The same plots are reproduced in Figure 8c,d when using model B. Now, an oscillatory behaviour on the BCWF and on the trajectories $X^j[t]$ is shown after time $t_s = 250$ fs. Such results can be understood by noticing that model B produces an increase in velocity in the Bohmian trajectories, but such faster Bohmian trajectories are not the *natural* behavior of the trajectories in the well when associated with only one eigenstate (they are expected to remain inside the well for a large time with a velocity close to zero). However, since the eigenstates of the quantum well form a complete basis, the mentioned oscillatory BCWF can be a solution to the Schrödinger equation there at the price of using many more eigenstates (with higher energies) to describe the new accelerated wave packet. Thus, the combination of several eigenstates in the well produces the oscillatory behaviour that we see in Figure 8d.

To better understand that model A provides a *natural* transition while model B provides an *unnatural* one, we show in Figure 9 the probability of the energy states $|c(E, t)|^2$ given by Equation (19) at $t = 0$ and $t = t_s^+$. The positive and negative energies only indicate scattering states injected from the left (positive) and injected from the right (negative). The blue line is the probability distribution of the energy eigenstates at the initial time $c(E, 0)$, while the red line is the same distribution but after scattering $c(E, t_s^+)$. In Figure 9a for model A, we observe a *natural* shift in the central energy given by $\langle E(t_s^+) \rangle = \langle E(t_s^-) \rangle + E_\gamma$, as expected. A definite argument in favor of model A (and against model B) is that the results in Figure 9a have an almost perfect agreement with the results in Figure 4b that were computed without approximation: the same transition happens from the first to the second energy eigenvalues of the quantum well. On the contrary, in Figure 9b for model B, a large amount of Hamiltonian eigenstates with negative energies (i.e., injection from the left) are populated after the scattering process. As explained, these additional energy components are the reason why we observe an oscillatory behaviour inside the well in Figure 8d. Model B is nonphysical because it does not satisfy the requirement of conservation of energy in the electron and photon collision. Since we deal with a wave packet (with some uncertainty on its energy), some deviation in the requirement of conservation of energy in each experiment is reasonable, but the deviations plotted in Figure 9b on the order of 1eV are not reasonable at all.

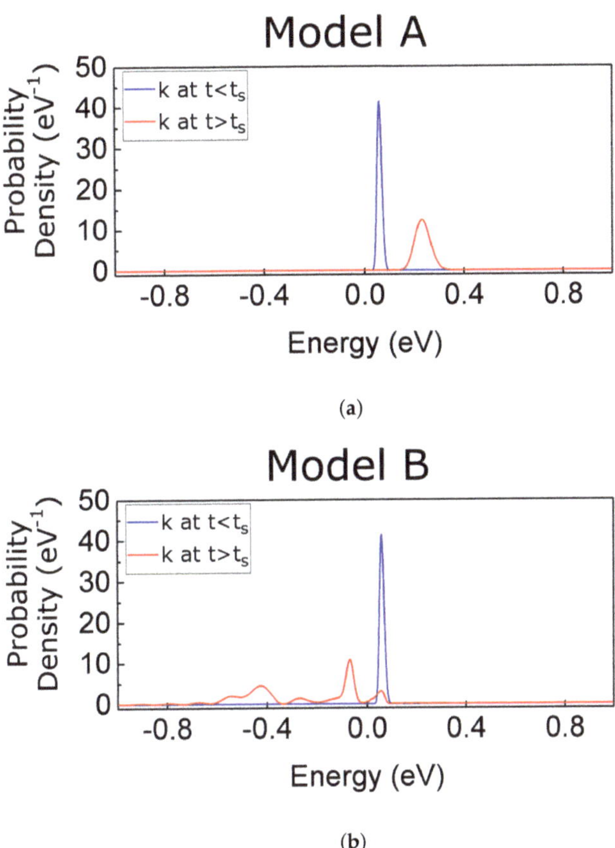

Figure 9. (**a**) Probability distribution of the Hamiltonian eigenstates for model A (spatial evolution shown in Figure 8b). (**b**) Probability distribution of the Hamiltonian eigenstates for model B (spatial evolution shown in Figure 8d). Blue lines represent the probability distribution of the Hamiltonian eigenstates before the scattering at $t < t_s$, while the red lines show it at $t > t_s$.

In conclusion, model B can only describe electron collisions when an approximation of flat potential is reasonable to describe the dynamic of the unperturbed electron. We get exactly the same conclusions when evaluating the emission process (not plotted) instead of the absorption process.

6. Conclusions

Quantum transport formalisms require modeling the perturbation induced by non-simulated degrees of freedom (like photons or phonons) on degrees of freedom of the simulated active region (the electrons). Among a number of different algorithms that allow us to include scattering events, here, we explore the possibility of implementing such scattering events as transitions between single-particle time-dependent pure states. We have shown that the Bohmian theory, through the use of BCWFs, allows for a rigorous implementation of transitions between pre- and post-selected single-particle pure states in the active device that is valid for both Markovian and non-Markovian conditions. Furthermore, we have shown that the practical implementation of such transitions requires one to model scattering events as a shift in central energies of BCWFs instead of a shift in central momenta. This last result seems to indicate dramatic consequences for quantum transport formalisms that introduce collisions through changes in momentum, e.g., the

Wigner function approach, when dealing with non-flat potential profiles where energy and momentum are non-commuting operators. This paper is part of a global and long-term research project that aims to develop the so-called BITLLES simulator [18]. We argue that the amount of information that this simulator framework can provide (from steady-state DC to transient and AC including the fluctuations of the current) in the quantum regime is comparable to the predicting capabilities of the traditional Monte Carlo solution of the Boltzmann transport equation in the semi-classical regime.

Author Contributions: Conceptualization, M.V., X.O., X.C., C.D. and G.A.; methodology, M.V., X.O., X.C., C.D. and G.A.; software, M.V. and X.O.; validation, M.V., X.O., X.C., C.D. and G.A.; investigation, M.V., X.O., X.C., C.D. and G.A.; writing–original draft preparation, M.V., X.O., X.C., C.D. and G.A.; writing—review and editing, M.V., X.O., X.C., C.D. and G.A.; visualization, M.V. and X.O.; supervision, X.O., X.C., G.A. and C.D.; project administration, X.O.; funding acquisition, X.O., G.A. and X.C. All authors have read and agreed to the published version of the manuscript.

Funding: This research was funded by Spain's Ministerio de Ciencia, Innovación y Universidades under grant No. RTI2018-097876-B-C21 (MCIU/AEI/FEDER, UE), by the "Generalitat de Catalunya" and FEDER for the project 001-P-001644 (QUANTUMCAT), by the European Union's Horizon 2020 research and innovation programme under grant No. 881603 GrapheneCore3, and by the Marie Skłodowska-Curie grant No. 765426 TeraApps.

Institutional Review Board Statement: Not applicable.

Informed Consent Statement: Not applicable.

Data Availability Statement: The data that support the findings of this study are available from the corresponding author, X.O., upon reasonable request.

Conflicts of Interest: The authors declare no conflict of interest. The funders had no role in the design of the study; in the collection, analyses, or interpretation of data; in the writing of the manuscript; or in the decision to publish the results.

Abbreviations

The following abbreviations are used in this manuscript:

BCWF	Bohmian Conditional Wave Function
BITLLES	Bohmian Interacting Transport in non-equiLibrium eLEctronic Structures
RTD	Resonant Tunneling Diode

References

1. Breuer, H.P.; Petruccione, F. *Theory of Open Quantum Systems*; Oxford University Press: Oxford, UK, 2002; pp. 5–23.
2. Klimeck, G. Single and multiband modeling of quantum electron transport through layered semiconductor devices. *J. Appl. Phys.* **1997**, *81*, 7845.
3. Klimeck, G.; Ahmed, S.S.; Bae, H.; Kharche, N.; Clark, S.; Haley, B.; Lee, S.; Naumov, M.; Ryu, H.; Saied, F.; et al. Atomistic Simulation of Realistically Sized Nanodevices Using NEMO 3-D Part I: Models and Benchmarks. *IEEE Trans. Electron Devices* **2007**, *54*, 2079–2089. [CrossRef]
4. Schmidt, A.; Cheng, B.; daLuz, M. Green function approach for general quantum graphs. *J. Phys. A Math. Gen.* **2003**, *36*, 42. [CrossRef]
5. Rossi, F. *Theory of Semiconductor Quantum Devices; The Density-Matrix Approach*; Springer: Berlin, Germany, 2010; pp. 89–130.
6. Iotti, C.; Ciancio, E.; Rossi, F. Quantum transport theory for semiconductor nanostructures: A density-matrix formulation. *Phys. Rev. B* **2005**, *72*, 125347. [CrossRef]
7. Wigner, E.P. On the quantum correction for thermodynamic equilibrium. *Phys. Rev.* **1932**, *40*, 749–759. [CrossRef]
8. Frensley, W. Wigner-Function Model of Resonant-Tunneling Semiconductor Device. *Phys. Rev. B* **1987**, *36*, 1570–1580. [CrossRef]
9. Weinbub, J.; Ferry, D.K. Recent advances in Wigner function approaches. *Appl. Phys. Rev.* **2018**, *5*, 041104. [CrossRef]
10. Querlioz, D.; Nguyen, H.-N.; Saint-Martin, J.; Bournel, A.; Galdin-Retailleau, S.; Dollfus, P. Wigner-Boltzmann Monte Carlo approach to nanodevice simulation: From quantum to semiclassical transport. *J. Comp. Electron.* **2009**, *8*, 324–335. [CrossRef]
11. Nedjalkov, M.; Querlioz, D.; Dollfus, P.; Kosina, H. Wigner Function Approach. In *Nano-Electronic Devices*; Springer: New York, NY, USA, 2011; pp. 289–358.
12. Fan, Z.; Garcia, J.; Cummings, A.; Barrios-Vargas, J.; Panhans, M.; Harju, A.; Ortmann, F.; Roche, S. Linear scaling quantum transport methodologies. *Phys. Rep.* **2021**, *903*, 1–69. [CrossRef]

13. Vyas, P.; Van de Put, M.; Fischetti, M. Master-Equation Study of Quantum Transport in Realistic Semiconductor Devices Including Electron-Phonon and Surface-Roughness Scattering. *Phys. Rev. Appl.* **2020**, *13*, 014067. [CrossRef]
14. Fischetti, M. Theory of electron transport in small semiconductor devices using the Pauli master equation. *J. Appl. Phys.* **1998**, *83*, 270. [CrossRef]
15. Kramer, T.; Kreisbeck, C.; Krueckl, V. Wave packet approach to transport in mesoscopic systems. *Phys. Scr.* **2010**, *82*, 038101. [CrossRef]
16. Bracher, C.; Delos, J.; Kanellopoulos, V.; Kleber, M.; Kramer, T. The photoelectric effect in external fields. *Phys. Lett. A* **2005**, *347*, 62–66. [CrossRef]
17. Bohm, D. A Suggested Interpretation of the Quantum Theory in Terms of "Hidden" Variables. *Phys. Rev.* **1952**, *85*, 166. [CrossRef]
18. Bohmian Interacting Transport in Non-equiLibrium eLEctronic Structures. Available online: europe.uab.es/bitlles (accessed on 29 March 2021).
19. Vacchini, B.; Smirne, A.; Laine, E.; Piilo, J.; Breuer, H. Markovianity and non-Markovianity in quantum and classical systems. *New J. Phys.* **2011**, *13*, 093004. [CrossRef]
20. Lindblad, G. On the generators of quantum dynamical semigroups. *Commun. Math. Phys.* **1976**, *48*, 119–130. [CrossRef]
21. Ferialdi, L. Exact Closed Master Equation for Gaussian Non-Markovian Dynamics. *Phys. Rev. Lett.* **2016**, *116*, 120402. [CrossRef] [PubMed]
22. Vega, I.; Alonso, D. Dynamics of non-Markovian open quantum systems. *Rev. Mod. Phys.* **2017**, *89*, 015001. [CrossRef]
23. Ghirardi, G.C.; Rimini, A.; Webber, T. Unified dynamics for microscopic and macroscopic systems. *Phys. Rev. D* **1986**, *34*, 470. [CrossRef]
24. Bassi, A.; Lochan, K.; Satin, S.; Singh, T.P.; Ulbricht, H. Models of wave-function collapse, underlying theories, and experimental tests. *Rev. Mod. Phys.* **2013**, *85*, 471. [CrossRef]
25. Strunz, W.T.; Diósi, L.; Gisin, N. Open System Dynamics with Non-Markovian Quantum Trajectories. *Phys. Rev. Lett.* **1999**, *82*, 1801. [CrossRef]
26. Strunz, W.T. The Brownian motion stochastic Schrödinger equation. *Chem. Phys.* **2001**, *268*, 237. [CrossRef]
27. Ferialdi, L.; Bassi, A. Exact Solution for a Non-Markovian Dissipative Quantum Dynamics. *Phys. Rev Lett.* **2012**, *108*, 170404. [CrossRef]
28. Gambetta, J.; Wiseman, H.M. Non-Markovian stochastic Schrödinger equations: Generalization to real-valued noise using quantum-measurement theory. *Phys. Rev. A* **2002**, *66*, 012108. [CrossRef]
29. Gambetta, J.; Wiseman, H.M. The interpretation of non-Markovian stochastic Schrödinger equations as a hidden-variable theory. *Phys. Rev. A* **2003**, *68*, 062104. [CrossRef]
30. Diósi, L.; Ferialdi, L. General Non-Markovian Structure of Gaussian Master and Stochastic Schrödinger Equations. *Phys. Rev. Lett.* **2014**, *113*, 200403. [CrossRef]
31. Oriols, X. Quantum-Trajectory Approach to Time-Dependent Transport in Mesoscopic Systems with Electron-Electron Interactions. *Phys. Rev. Lett.* **2007**, *98*, 066803. [CrossRef] [PubMed]
32. Dürr, D.; Teufel, S. *Bohmian Mechanics: The Physics and Mathematics of Quantum Theory*; Springer: Berlin, Germany, 2009.
33. Oriols, X.; Mompart, J. *Applied Bohmian Mechanics: From Nanoscale Systems to Cosmology*, 2nd ed.; Jenny Stanford Publishing: Singapore, 2019.
34. Colomés, E.; Zhan, Z.; Marian, D.; Oriols, X. Quantum dissipation with conditional wave functions: Application to the realistic simulation of nanoscale electron devices. *Phys. Rev. B* **2017**, *96*, 075135. [CrossRef]
35. Albareda, G.; López, H.; Cartoixà, X.; Suñé, J.; Oriols, X. Time-dependent boundary conditions with lead-sample Coulomb correlations: Application to classical and quantum nanoscale electron device simulators. *Phys. Rev. B* **2010**, *82*, 085301. [CrossRef]
36. Marian, D.; Colomés, E.; Oriols, X. Quantum noise from a Bohmian perspective: Fundamental understanding and practical computation in electron devices. *J. Phys. Condens. Matter* **2015**, *27*, 245302. [CrossRef]
37. Albareda, G.; Traversa, F.L.; Benali, A.; Oriols, X. Computation of Quantum Electrical Currents throught The Ramo–Shockley–Pellegrini Theorem with Trajectories. *Fluctuation Noise Lett.* **2012**, *11*, 1242008. [CrossRef]
38. Jacoboni, C.; Reggiani, L. The Monte Carlo method for the solution of charge transport in semiconductors with applications to covalent materials. *Rev. Mod. Phys.* **1983**, *55*, 645. [CrossRef]
39. Tzemos, A.C.; Contopoulos, G.; Efthymiopoulos, C. Bohmian trajectories in an entangled two-qubit system. *Phys. Scr.* **2019**, *94*, 105218. [CrossRef]
40. Tzemos, A.C.; Contopoulos, G. Chaos and ergodicity in an entangled two-qubit Bohmian system. *Phys. Scr.* **2020**, *99*, 065225.

Article

Two-Excitation Routing via Linear Quantum Channels

Tony John George Apollaro [1,*,†] and Wayne Jordan Chetcuti [2,3,4,†]

1   Department of Physics, Faculty of Science, University of Malta, MSD 2080 Msida, Malta
2   Dipartimento di Fisica e Astronomia, Via S. Sofia 64, 95127 Catania, Italy; wayne.chetcuti@dfa.unict.it
3   INFN-Sezione di Catania, Via S. Sofia 64, 95127 Catania, Italy
4   Quantum Research Centre, Technology Innovation Institute, Abu Dhabi 9639, UAE
*   Correspondence: tony.apollaro@um.edu.mt; Tel.: +356-2340-8101
†   These authors contributed equally to this work.

Abstract: Routing quantum information among different nodes in a network is a fundamental prerequisite for a quantum internet. While single-qubit routing has been largely addressed, many-qubit routing protocols have not been intensively investigated so far. Building on a recently proposed many-excitation transfer protocol, we apply the perturbative transfer scheme to a two-excitation routing protocol on a network where multiple two-receivers block are coupled to a linear chain. We address both the case of switchable and permanent couplings between the receivers and the chain. We find that the protocol allows for efficient two-excitation routing on a fermionic network, although for a spin-$\frac{1}{2}$ network only a limited region of the network is suitable for high-quality routing.

Keywords: quantum state routing; many-body dynamics; quantum information; fermionic network

1. Introduction

The coherent transfer of excitations from a sender to a receiver, located at different positions in a network, is of primary importance for many quantum-based technological applications, ranging from spintronics and atomtronics [1] to quantum-information processing [2].

While a great amount of work has been devoted to the routing of the quantum state of a single qubit [3–11], where the fidelity of the transfer protocol can be expressed in terms of the transition amplitude of a single excitation between a sender and a receiver location [12], the routing of a multiple qubit state is a far less investigated scenario. Although several protocols have been proposed both for two-qubit and multi-partite entangled quantum state transfer [13–22], their extension to a routing configuration on an arbitrary network is not straightforward. One reason being that almost all the proposed protocols rely on the quantum channel possessing mirror-symmetry, which, allowing for multiple receivers at arbitrary positions, is difficult to attain: in Ref. [23] it has been shown, e.g., that perfect state routing between multiple sites with real Hamiltonians is impossible. Moreover, the presence of a sender and a receiver block located at positions other than the edges of a 1D quantum channel, implies that the total system is no longer one-dimensional and the fermionisation of the spin chain via the celebrated Jordan–Wigner mapping is not valid anymore [24]. As a consequence, the full spectrum of the network's Hamiltonian has to be found in the Hilbert space sector with two excitations and this can become, for long chains, quite cumbersome.

In this work we investigate the routing of two excitations by means of a linear chain, acting as a quantum wire, to which receivers can connect at arbitrary positions. Following the results of our recent work [25], we apply the weak-coupling protocol in order to route fermionic excitations on a 2D network. We consider both the case of switchable and permanent couplings of the receiver block to the quantum wire, obtaining the receivers' locations which allow for perturbatively perfect two-excitation transfer. We then compare the routing performance of fermions, which due to the non-interacting nature of the Hamiltonian considered in our work, can be analyzed in terms of single-particle transition

Entropy **2021**, *23*, 51. https://doi.org/10.3390/e23010051   https://www.mdpi.com/journal/entropy

amplitudes, to the case where the network hosts spin-$\frac{1}{2}$ particles interacting via the XX-Heisenberg type Hamiltonian. We find that, although, a rigorous mapping of spins to non-interacting fermions is not possible because of the 2D nature of the network, it is indicated that several features of the free fermions dynamics can be retrieved also in the spin dynamics.

The paper is organised as follows. In Section 2, a brief introduction to the many-body dynamics in non-interacting fermion systems on a discrete lattice is given; in Section 3, the proposed protocol of two-excitation routing, both with switchable and permanent couplings, on a 2D lattice is presented; in Section 4, we analyze the case of spin-$\frac{1}{2}$ particle occupying the lattice positions of the network. Finally, in Section 5 we discuss the main findings of our research and outline some future directions.

2. Many-Body Dynamics in Non-Interacting Fermions on a Discrete Lattice

Let us consider a discrete lattice model where each site can host one spinless fermion and whose dynamics is governed by the hopping Hamiltonian

$$\hat{H} = \sum_{\langle ij \rangle} J_{ij} \left(\hat{c}_i^\dagger \hat{c}_j + h.c \right) , \tag{1}$$

where $\hat{c}_i^\dagger$ ($\hat{c}_i$) is the creation (annihilation) operator of a fermion on site i and J_{ij} is the kinetic term accounting for the hopping of a fermion between neighboring sites i and j. This Hamiltonian conserves the total number of excitations (fermions) and can be block-diagonalised in each fixed particle-number sector. Moreover, because of the quadratic nature of the Hamiltonian, only the spectrum in the single-excitation subspace is needed in order to retrieve the full energy spectrum. This is a consequence of the non-interacting nature of the Hamiltonian witnessed by the absence of quartic terms accounting for particle-particle interactions [26]. We report here, for the sake of completeness, the mains steps for the derivation of the many-body dynamics in terms of single-body dynamics for non-interacting fermions, which is standard procedure in the second-quantization formalism.

The diagonalized form of the Hamiltonian in Equation (1) in the single-excitation sector reads

$$\hat{H} = \sum_{k=1}^{N} E_k |E_k \rangle \langle E_k| = E_k \hat{c}_k^\dagger \hat{c}_k \tag{2}$$

where $\{E_k, |E_k\rangle\}$ are the eigenvalues and eigenvectors of the N-dimensional adjacency matrix of the graph with entries J_{ij}. Expressed in the position basis, $|n\rangle \equiv \hat{c}_n^\dagger |0\rangle = |00\ldots1_n00\ldots\rangle$, where $|0\rangle$ represents the fermionic vacuum state and $|1_n\rangle$ denotes the presence of a fermion on site n, the energy eigenstates in the single-excitation sector read $|E_k\rangle = \sum_{n=1}^{N} a_{kn}|n\rangle$, with $a_{kn} = \langle n|E_k\rangle$. The single-particle transition amplitude of an excitation from site s to site r is given by

$$f_s^r(t) = \langle r|e^{-i\hat{H}t}|s\rangle = \sum_{k=1}^{N} a_{r,k} a_{s,k}^* e^{-iE_k t} . \tag{3}$$

Because of the non-interacting nature of the Hamiltonian in Equation (1), the energy eigenstates in the Hilbert space with m fermionic excitations are given by

$$\left|E_{k_1 k_2 \ldots k_m}\right\rangle = \sum_{n_1 < n_2 < \cdots < n_m = 1}^{N} a_{n_1 n_2 \ldots n_m, k_1 k_2 \ldots k_m} |n_1 n_2 \ldots n_m\rangle , \tag{4}$$

with eigenvalues $E_{k_1 k_2 \ldots k_m} = E_{k_1} + E_{k_2} + \cdots + E_{k_m}$ and $a_{n_1 n_2 \ldots n_m, k_1 k_2 \ldots k_m}$ denoting the Slater determinant.

The many-body transition amplitude of m excitations from sites $\mathbf{s} = \{s_1, s_2, \ldots, s_m\}$ to sites $\mathbf{r} = \{r_1, r_2, \ldots, r_m\}$ is readily obtained as a determinant of a matrix whose entries are the single-particle transition amplitudes in Equation (3),

$$
\begin{aligned}
f_{\mathbf{s}}^{\mathbf{r}}(t) = \langle \mathbf{r} | e^{-i\hat{H}t} | \mathbf{s} \rangle &= \sum_{k_1 < k_2 < \cdots < k_m = 1}^{N} e^{-i\left(E_{k_1} + E_{k_2} + \cdots + E_{k_m}\right)t} \langle r_1 r_2 \ldots r_m | E_{k_1 k_2 \ldots k_m} \rangle \langle E_{k_1 k_2 \ldots k_m} | s_1 s_2 \ldots s_m \rangle \\
&= \begin{vmatrix}
f_{s_1}^{r_1}(t) & f_{s_1}^{r_2}(t) & \cdots & f_{s_1}^{r_m}(t) \\
f_{s_2}^{r_1}(t) & \cdots & \cdots & f_{s_2}^{r_m}(t) \\
\vdots & & \ddots & \vdots \\
f_{s_m}^{r_1}(t) & & \cdots & f_{s_m}^{r_m}(t)
\end{vmatrix} .
\end{aligned}
\tag{5}
$$

The expression given in Equation (5) holds for every fermionic quadratic model, whereas if the operators in Equation (1) represent bosons, then, instead of the determinant, the many-body transition amplitude is given by the permanent of the matrix [25,26].

3. The Model

In this section, we apply the formalism of Section 2 to determine the two-excitation transition probability from a sender block to a receiver block, both composed of two sites, that are connected to a linear chain. The aim is to derive the conditions for the routing of the two excitations from the senders' to the receivers' location. We will analyze two networks: (a) the receiver blocks have switchable couplings to the wire (Figure 1); (b) the receiver blocks are permanently coupled to the wire and the hopping term J_s in the sender block is tunable (Figure 3).

We consider Hamiltonians of the type given in Equation (1), which, decomposed into the different components of the network, i.e., sender S, wire W, and receivers R, read

$$
\hat{H} = \hat{H}_S + \sum_i \hat{H}_{R_i} + \hat{H}_W + \hat{H}_{SW} + \sum_i \hat{H}_{R_i W} .
\tag{6}
$$

The Hamiltonian of the sender block and the i-th receiver block are, respectively

$$
\hat{H}_S = J_s \left(\hat{c}_1^\dagger \hat{c}_2 + h.c. \right), \quad \hat{H}_{R_i} = J_i \left(\hat{c}_{r_i}^\dagger \hat{c}_{r_i+1} + h.c. \right),
\tag{7}
$$

with r_i denoting the position on the graph which will be given in the following. The Hamiltonian for the quantum data bus reads

$$
\hat{H}_w = J \sum_{n=1}^{n_w - 1} \left(\hat{c}_n^\dagger \hat{c}_{n+1} + h.c. \right) .
\tag{8}
$$

where n_w denotes the length of the wire. Finally, the coupling between the sender block and the data bus site is assumed to be in the weak-coupling regime, $J_0 \ll J, J_s, J_i$

$$
\hat{H}_{Sw} = J_0 \left(\hat{c}_2^\dagger \hat{c}_3 + h.c. \right) ;
\tag{9}
$$

as well as the coupling between the i-th receiver block at location r_i and the corresponding data bus site w_i, where $1 \leq w_i \leq n_w$

$$
\hat{H}_{R_i w} = J_0 \left(\hat{c}_{r_i}^\dagger \hat{c}_{w_i} + h.c. \right) .
\tag{10}
$$

For case (a) all couplings between the receiver blocks and the wire are switched off but one, embodying the recipient of the routing protocol and we set $r_i = n_w + 2$; see Figure 1 for an instance of the numbering choice of the sites following the sender-wire-receiver order; the same ordering is followed for case (b).

The whole system sender+wire+receivers is made up of $N = n_w + 2(n_r + 1)$ sites with r denoting the number of receiver blocks.

From Equation (5), the two-body transition probability, with $\mathbf{s} = \{s_1, s_2\}$ and $\mathbf{r_i} = \{r_{i,1}, r_{i,2}\}$, is given by

$$|f_{\mathbf{s}}^{\mathbf{r_i}}(t)|^2 = \left|\langle 1,2|e^{-it\hat{H}}|N-1,N\rangle\right|^2 = \begin{vmatrix} f_1^{N-1}(t) & f_1^{N}(t) \\ f_2^{N-1}(t) & f_2^{N}(t) \end{vmatrix}^2 . \tag{11}$$

For only one sender and one receiver block located at opposite edges of the quantum wire, the model is one-dimensional and, using the Jordan–Wigner mapping from spinless fermions to spin-$\frac{1}{2}$ particles, the Hamiltonian in Equation (6) with open boundary conditions is equivalent to the XX spin-$\frac{1}{2}$ model with nearest-neighbor coupling.

$$\hat{H} = \sum_{n=1}^{N} \frac{J_n}{2} \left(\hat{\sigma}_n^x \hat{\sigma}_{n+1}^x + \hat{\sigma}_n^y \hat{\sigma}_{n+1}^y \right) . \tag{12}$$

In such a case, it has been shown that two-qubit quantum state transfer [13–15] as well as entanglement generation of two Bell states [27] is achieved with high fidelity. Modifications of the one-dimensional geometry have been investigated too. In Refs. [17,18] each spin of the sender (receiver) block is coupled to the edges of the 1D quantum wire allowing for the transfer of a Bell state when operating in the single-excitation subspace. A similar geometry is adopted in Refs. [4,6] with multiple sender (receiver) non-interacting spins coupled to the wire at the edges.

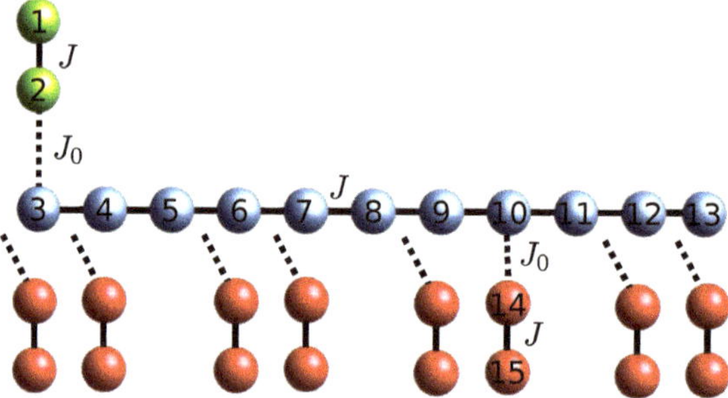

Figure 1. Quantum routing of excitations by means of a linear chain quantum data bus with switchable interactions. The sender and receiver sites are depicted in green and red, respectively, while the quantum data bus sites are in blue. Continuous lines represent permanent couplings $J = 1$, while dotted lines encode switchable weak couplings $J_0 \ll 1$; also shown is the numbering choice of the sites position adopted in Section 3.1.

3.1. Routing with Switchable Weak Couplings

Here we consider only one receiver block coupled to the wire for each execution of the routing protocol and, as we shall see, this allows us to assume uniform coupling within each component of the setup, i.e, the sender, the wire and the receiver blocks. We choose $J_s = J = J_r = 1$ as our energy and time unit. On the other hand, the couplings between sender (receiver) block and the wire will be in the weak-coupling regime, which we set throughout the paper to $J_0 = 0.01$.

The 1D-case where only one block of senders and one of receivers is each coupled at the edge of the quantum wire has been addressed in Ref. [25]. There it has been shown that, although each length of the quantum wire n_w allows for high-fidelity excitation transfer, for $n_w = 3l + 2$ ($l = 0, 1, 2, \dots$), resonances between the sender (receiver) and the wire single-particle energy levels give rise to a faster transfer with respect to the instances $n_w = 3l, 3l + 1$ where off-resonant transfer takes place. In the former case, the single-particle transfer occurs on a time scale of the order of J_0^{-1}, yielding to a two-excitation transfer time scale of the order of $10\, J_0^{-1}$ with the reason for the multiplicative factor being that the transfer dynamics involves a difference between eigenenergies that are perturbed to first-order in J_0. On the other hand, for the off-resonant dynamics, the two-particle transfer time is of order J_0^{-2}. Considering that the excitation transfer mechanism holds in the perturbative regime $J_0 \ll 1$, this may translate in severals of magnitude.

Here we address the case where the receiver block is coupled to the quantum wire at a different position w_i than the edge of the chain; see Figure 1. We also omit the suffix i since only one receiver block is present in this protocol. We aim at finding the conditions on the position w for which resonant transfer of the excitations from the sender to the receiver block at takes place. Following the argument for faster (resonant) transfer in Ref. [25], we set the length of the quantum wire $n_w = 3l + 2$ so that perturbative transfer is achieved for $w = n_w$, i.e., a receiver block can be coupled to the edge of the wire. For this wire length, we find that it is possible to couple a receiver block at each site $w \neq 3p$ of n_w, with p integer. The fact that these latter sites of the wire cannot act as connection points for the receiver block can be explained by looking at the eigenstates of the wire's Hamiltonian $\hat{H}_W$ (Equation (8)) that are resonant with the eigenstates of the sender (receiver) block. For $J = 1$, the unperturbed energy level of the sender (receiver) is $E_{res} = \pm 1$, and, because of the mirror-symmetry of the Hamiltonians in Equation (7), they have the identical (absolute value) overlap on each site [28] so that it suffices to consider only one of them. The unperturbed energy levels of the wire that are resonant with the sender (receiver) are given by $E_k = 2\cos\frac{k\pi}{n_w+1} = 1$. Therefore, we obtain that, ordering $E_k's$ in decreasing order, the $k = \frac{n_w+1}{3}$-energy level of the wire is the resonant one, see left panel of Figure 2 for a schematic representation of the resonance condition. Expressing the corresponding energy eigenstate in the position basis

$$|E_{res}\rangle = \sqrt{\frac{2}{n_w+1}}\sum_{m=1}^{n_w}\sin\frac{km\pi}{n_w+1}|m\rangle = \sqrt{\frac{2}{n_w+1}}\sum_{m=1}^{n_w}\sin\frac{m\pi}{3}|m\rangle, \tag{13}$$

meaning that the resonant energy level has no support on any site of the wire being multiple of 3. As a consequence, at first-order perturbation theory, the resonant energy level does not overlap with the receiver sites coupled to each third site of the wire, making the latter not apt as connection points for a two-excitation transfer. Furthermore, the k_{res} eigenenergy state has constant spatial overlap with every other site $m \neq 3p$. This translates into a symmetric spatial distribution of the first-order perturbed eigenstates on the sender and receiver block, thus enabling the excitation transfer. Hence, for a wire of length $n_w = 3l + 2$ with uniform couplings equal to those within the sender (receiver) block, a total of $n_r = 2(l+1)$ receiver points are possible. In Figure 2 an instance of such a protocol is shown for $l = 3$ with the receiver pair n_{r_6} coupled to the quantum wire. In the right panel of Figure 2 an instance of the Rabi-like oscillations are shown for a wire's length of $n_w = 11$ and connection point of the receiver block at $w = 7$. We found in our numerical simulations for lengths of the wire in the order of the hundreds, that also for longer chains the fidelity reaches $F = 1 - O(J_0)$ for a receiver block connected at $w \neq 3p$ with the first peak of the oscillations occurring at a time of order $10\,J_0^{-1}$.

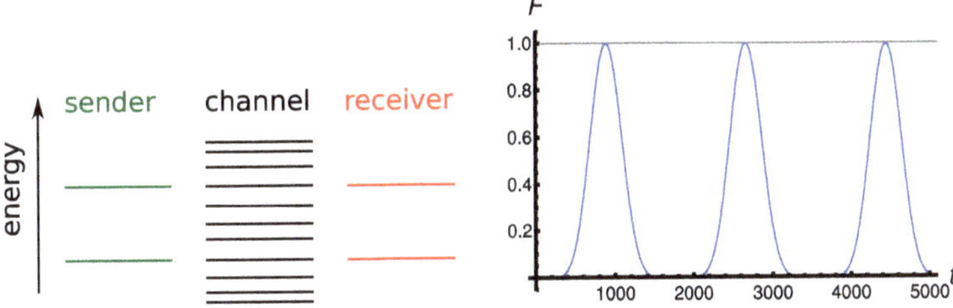

Figure 2. (**left**) Single-excitation energy levels in the switchable coupling configuration. (**right**) Two-excitation transfer fidelity in the switchable configuration of Section 3.1 with $n_w = 11$, $r = 7$, and $J_0 = 0.01$.

3.2. Routing with Permanent Weak Couplings

A much more desirable routing configuration would be one without the need of switching on and off the couplings as described in the previous section. For the routing of a single particle, this has been achieved in Ref. [3] where in both the linear and the circular geometry, the sender chooses the receiver site tuning the only single-energy level on resonance with the receiver (and the quantum wire) energy level by means of a local magnetic field, i.e., the value of the local magnetic field acting on the receiver qubit is the routing address. In the case of a sender block embodied by two particles, the same strategy does not work as a local magnetic field produces an uniform shift of both of the two energy levels and the simple sinusoidal excitation dynamics is lost. However, it is still possible to perform resonant routing with a sender block of two sites by using the intraspin coupling, which results in a symmetric energy shrinking or dilatation of the two single-energy levels. In such a case, the routing address of each receiver block is given by their intraspin coupling J_r; see Figure 3 for the geometry of the network and the left panel of Figure 4 for the single-excitation energy levels.

In Section 3.1, we have shown that, for $J_s = 1$, the wire's $k = \frac{n_w+1}{3}$-energy level is resonant with the sender block, and allows for the transfer of the two excitations to a receiver block with intraspin coupling $J_r = J_s$ provided that the connection point along the wire is $w \neq 3p$. The very same argument can be applied by tuning J_s to a different value so that the single-energy levels of the sender block $E_s = \pm 2J_s$ are resonant with two (symmetric) energy levels of the wire. In order to match the resonance condition, an integer solution for k has to satisfy the following equation

$$J_s = \cos \frac{k\pi}{n_w + 1}. \tag{14}$$

That is, the $k = \frac{n_w+1}{\pi} \arccos J_s$-th energy eigenvalue of the wire is resonant with the sender. For example, for $J_s = \frac{\sqrt{3}}{2}$, $k = \frac{n_w+1}{6}$. Hence, the allowed contact points w_i along the wire have to fulfill the condition that the resonant eigenstate spatial component of the contact point of the sender has to be equal to that of the receiver's contact point, i.e.,

$$\sqrt{\frac{2}{n_w + 1}} \sin \frac{k_{res} s \pi}{n_w + 1} = \sqrt{\frac{2}{n_w + 1}} \sin \frac{k_{res} w_i \pi}{n_w + 1}. \tag{15}$$

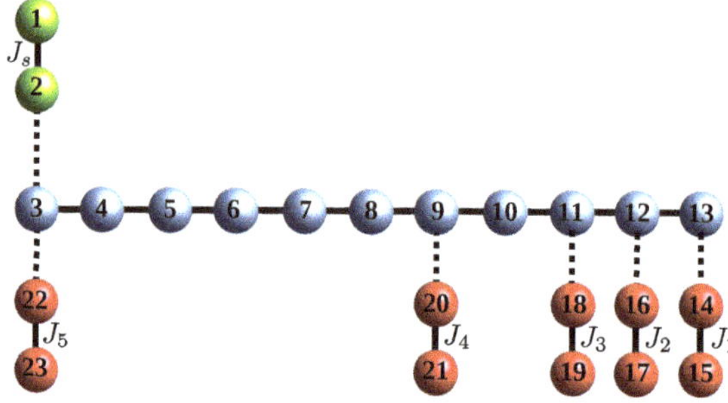

Figure 3. Quantum routing of excitations by means of a linear chain quantum data bus with permanent interactions and assuming J_s tuneable. The sender and receiver sites are depicted in green and red, respectively, while the quantum data bus sites are in blue; also shown is the numbering choice of the sites position adopted in Section 3.2.

To conclude this section, we recap the main results for routing a pair of excitations across a wire with uniform couplings to a desired location, specifying the resonance conditions on the sender and receiver couplings, respectively J_s and J_{r_i}, and the respective allowed contact points.

For a wire of length n_w, the possible communication parties are $\frac{n_w}{2}$ for even length chain and $\frac{n_w-1}{2}$ for odd length ones. Setting the intrawire coupling $J = 1$, the single energy levels for which first-order excitation transfer occurs are $E_k = 2\cos\frac{k\pi}{n_w+1}$, $k = 1, 2, \ldots, \frac{n_w}{2}$ ($k = 1, 2, \ldots, \frac{n_w-1}{2}$ for odd length chains). Each k determines the intraspin coupling of the sender block via the relation $J_{r_i} = 2\cos\frac{k_i\pi}{n_w+1}$ and the possible contact points w_i along the wire via

$$\sin\frac{k_i s\pi}{n_w + 1} = \sin\frac{k_i w_i \pi}{n_w + 1}. \tag{16}$$

Assuming that the sender is attached to the first site of the wire $s = 1$, and exploiting the periodicity of the sin function, $|\sin\alpha| = |\sin(\alpha \pm n\pi)|$, with n integer,

$$\frac{k_i w_i}{n_w + 1} = \frac{k_i\pi}{n_w + 1} \pm n\pi \rightarrow w_i = \left|1 \pm \frac{n(n_w + 1)}{k_i}\right|. \tag{17}$$

Finally, the allowed contact points w_i are the integers $\in [1, n_w]$ satisfying Equation (17). An instance of these conditions is given in Figure 4 for $n_w = 11$ and the corresponding values of J_{r_i} and r_i are given in Table 1.

Table 1. Values of the intraspin couplings for the receiver blocks and available wire's connection sites for the receiver block for a wire's length $n_w = 11$.

k	J_r	w_i
1	$\frac{\sqrt{3}-1}{2}$	1,11
2	1	1,2,4,5,7,8,10,11
3	$\sqrt{2}$	1,3,5,7,9,11
4	$\sqrt{3}$	1,5,7,11
5	$\frac{\sqrt{3}+1}{2}$	1,11

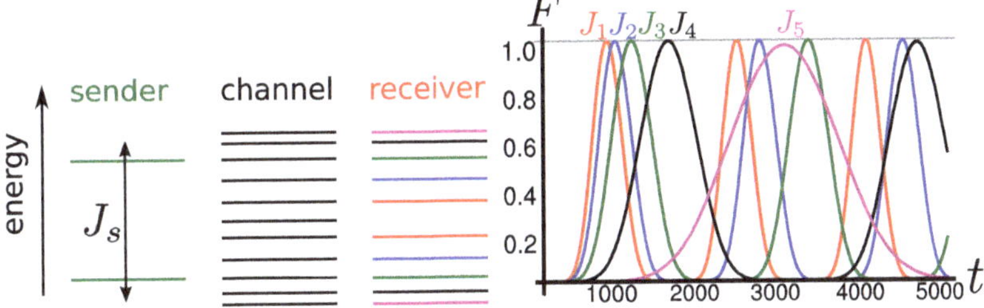

Figure 4. (**left**) Single-excitation energy levels in the permanently coupled routing scheme. The sender's energy level can be tuned to be in resonance with a different pair of wire's (and receiver's) energy levels by tuning J_s. (**right**) Excitation transfer in the permanent coupling configuration of Section 3.2 with $n_w = 11$, $J_0 = 0.01$ and coupling scheme as in Figure 3. The different curves correspond to the transfer fidelity of the two excitation to different receiver block by tuning J_s to J_{r_i}. The colors of the curves correspond to the enegy levels in the left panel.

4. Routing in Spin Systems

In the previous Sections, we have shown how, in the weak-coupling regime, routing of two-excitations from a sender to a receiver block can be achieved both in a switchable and a permanent coupling configuration in quadratic Hamiltonians. In this Section, we will consider the case when the network is made up of spin-$\frac{1}{2}$ particles interacting via an

XX-type Heisenberg Hamiltonian. We will consider the switchable routing configuration depicted in Figure 1 with Hamiltonian

$$\hat{H} = \sum_{\langle ij \rangle} \frac{J_{ij}}{2} \left(\hat{\sigma}_i^x \hat{\sigma}_j^x + \hat{\sigma}_i^y \hat{\sigma}_j^y \right), \tag{18}$$

where $\langle \rangle$ denotes the summation running over nearest-neighbor sites. Notice that Equation (18) differs from Equation (12) because of the 2D nature of the network. By introducing the ladder operators $\hat{\sigma}^{\pm} = \frac{\hat{\sigma}^x \pm i \hat{\sigma}^y}{2}$, the Hamiltonian of the system can be obtained from Equations (6)–(10) by substitution of $\hat{c} \to \hat{\sigma}^-$ and $\hat{c}^\dagger \to \hat{\sigma}^+$. As already stated in Section 3, were the network one-dimensional, i.e., the receiver block coupled to the last spin of the wire, then the Jordan–Wigner transformation would map Equation (18) to a quadratic spinless fermion Hamiltonian. Such a case would constitute a special instance of the analysis in Section 3.1 and several works on two-qubit quantum state transfer can be found in the literature. However, in the general case, where the receiver block is coupled to an arbitrary site of the wire, the Jordan–Wigner mapping does not apply as the system looses its one-dimensional nature.

However, for a configuration such as the one depicted in Figure 1, there is only one spin belonging to the wire that has three nearest-neighboring spins; therefore, the one-dimensional nature of the model is only locally broken with the lowest possible coordination number. It is therefore interesting to investigate if the routing properties of the spinless non-interacting one-dimensional model of Section 3 still persist also when the network is made of spins when a rigorous mapping to fermions is not possible.

Now, in order to evaluate the transition probability of two-excitations, we need to diagonalise the Hamiltonian in Equation (18) in the two-excitations sectors, being the reduction to one-particle transition amplitudes not possible. The dimension of the Hamiltonian in the two-excitation Hilbert space is the binomial factor $\dim\left[\hat{H}^{(2)}\right] = \binom{N}{2}$ and we diagonalise the Hamiltonian numerically for $N = 306$ using the QuSpin package [29].

From Figure 5 we see that, as for the free-fermion network in Section 3.1, the transition probability of the two excitations from the sender block to the receiver block is negligible whenever the latter is coupled to every third spin of the linear chain. Furthermore, high-quality two-excitation transfer can be achieved, on a time-scale similar to that of the free-fermion network, only if the receiver block is coupled to connection points of the wire at the opposite edge with respect to the sender block. Moving away from that edge causes a linear decrease of the quality of the transfer with a lower slope the longer the wire. This may be seen as a consequence of the fact that the longer the wire, the more the one-dimensional nature of the system becomes manifest.

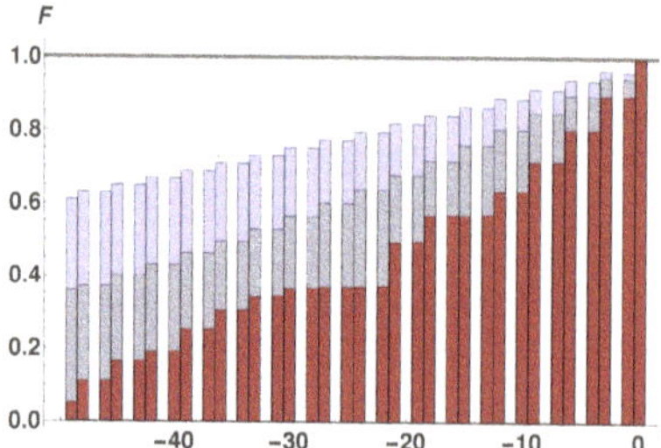

Figure 5. Transition probability for the switchable couplings protocol in Section 3.1 with spin-$\frac{1}{2}$ particles sitting on the graph with $N = 78$ (red), $N = 156$ (gray), and $N = 306$ (blue) and interacting via the *XX* Hamiltonian in Equation (18). The numbering on the *x*-axis is the distance from the edge opposite the sender block $n = 0, -1, -2, \ldots$. Notice that, for the receiver block coupled to each third n of the wire, the transiton probability is negligible. Interestingly, the quality of the transfer increases with the wire's length.

5. Discussion

In this paper, we have investigated the routing of two fermionic excitations across a quantum network. In the proposed protocol we were able to show that two fermions, initially located on a sender block composed of two sites, can be efficiently routed to a receiver block of two sites, provided that both the former and the latter are weakly coupled to a one-dimensional quantum wire, modeled by a fermionic nearest-neighbor hopping Hamiltonian. We have proposed two different protocols: in the first one, we have assumed switchable couplings and derived the connection points of the wire which yields high-quality routing; in the second one we have assumed permanent couplings and envisaged in the tunability of the sender's intrasite coupling a mean to route the two excitations to the desired location. In each considered configuration we obtained a perturbatively-perfect fidelty, i.e., $F = 1 - O(J_0)$, where J_0 is the weak coupling of the sender and receiver block to the wire, with a transfer time scaling as $O(10J_0^{-1})$. We also compared the fermionic network with a spin-$\frac{1}{2}$ network interacting via an XX-Heisenberg Hamiltonian. Due to the 2D nature of the network, the analysis had to rely on numerical evaluation and we found that, apart from the scenario where the receiver blocks are located towards the end of the wire, efficient routing is not achievable with qubits. However, our work hints towards the possibility to utilise very long quantum wires for the proposed 2-qubit routing protocol as we observed an enhancement of the routing fidelity by increasing the length of the wire. In such a scenario, our protocol may be utilised also for two-qubit entanglement routing, similarly to Refs. [17,18] where the tranfer is achieved between the edges of the chain.

For a realistic implementation of our protocol one should however consider possible experimental imperfections in the couplings, the time-dependence of the sender/receiver couplings to the chain, and decoherence due to interaction with the surrounding environment. While an extensive analysis of the performance of our protocol under these conditions has not been presented here, and may be left to future investigation, it is reasonable to assume that similar analyses done for the one-particle transfer scenario may apply also for two-particle routing as our results are derived from single-particle transition amplitudes. In this regard, disorder in couplings [21,30–33], time-dependent couplings [34,35], and decoherence [36,37] have been extensively addressed and several strategies to counter the detrimental effect on the transfer quality have been devised which may find application also in our two-particle routing protocol.

Author Contributions: Conceptualization, T.J.G.A. and W.J.C.; Data curation, T.J.G.A. and W.J.C.; Software, T.J.G.A. and W.J.C.; Writing—original draft, T.J.G.A. and W.J.C. All authors have read and agreed to the published version of the manuscript.

Funding: This research has been carried out using computational facilities procured through the European Regional Development Fund, Project No. ERDF-080 A supercomputing laboratory for the University of Malta.

Data Availability Statement: The data presented in this study are available on request from the corresponding author.

Acknowledgments: We thank Kristian Grixti and Alessio Magro for technical support with the computational facilities.

Conflicts of Interest: The authors declare no conflict of interest.

References

1. Amico, L.; Boshier, M.; Birkl, G.; Minguzzi, A.; Miniatura, C.; Kwek, L.C.; Aghamalyan, D.; Ahufinger, V.; Andrei, N.; Arnold, A.S.; et al. Roadmap on Atomtronics. 2020. Available Online: https://hal.archives-ouvertes.fr/hal-02933999/ (accessed on 30 December 2020)
2. Nikolopoulos, G.M.; Jex, I. *Quantum State Transfer and Network Engineering*; Springer: Berlin, Germany, 2014; pp. 1–250. [CrossRef]
3. Paganelli, S.; Lorenzo, S.; Apollaro, T.J.G.; Plastina, F.; Giorgi, G.L. Routing quantum information in spin chains. *Phys. Rev. A* **2013**, *87*, 062309. [CrossRef]
4. Yousefjani, R.; Bayat, A. Simultaneous multiple-user quantum communication across a spin-chain channel. *Phys. Rev. A* **2020**, *102*, 012418. [CrossRef]

5.   Bayat, A.; Bose, S.; Sodano, P. Entanglement Routers Using Macroscopic Singlets. *Phys. Rev. Lett.* **2010**, *105*, 187204. [CrossRef] [PubMed]

6.   Yousefjani, R.; Bayat, A. Parallel Entangling Gate Operations and Two-Way Quantum Communication in Spin Chains. *arXiv* **2020**, arXiv:2008.12771. Available Online: https://arxiv.org/abs/2008.12771 (accessed on 30 December 2020)

7.   Zueco, D.; Galve, F.; Kohler, S.; Hänggi, P. Quantum router based on ac control of qubit chains 1. *Phys. Rev. A* **2009**, 1–10. [CrossRef]

8.   Chen, B.; He, Y.Z.; Chu, T.T.; Shen, Q.H.; Zhang, J.M.; Peng, Y.D. Efficient routing quantum information in one-dimensional tight-binding array. *Prog. Theor. Exp. Phys.* **2020**, *2020*, 1–11. [CrossRef]

9.   Wanisch, D.; Fritzsche, S. Driven spin chains as high-quality quantum routers. *Phys. Rev. A* **2020**, *102*, 032624. [CrossRef]

10.   Pemberton-Ross, P.J.; Kay, A. Perfect Quantum Routing in Regular Spin Networks. *Phys. Rev. Lett.* **2011**, *106*, 020503. [CrossRef]

11.   Zhan, X.; Qin, H.; Bian, Z.H.; Li, J.; Xue, P. Perfect state transfer and efficient quantum routing: A discrete-time quantum-walk approach. *Phys. Rev. A* **2014**, *90*, 012331. [CrossRef]

12.   Bose, S. Quantum Communication through an Unmodulated Spin Chain. *Phys. Rev. Lett.* **2003**, *91*, 207901. [CrossRef]

13.   Apollaro, T.J.G.; Lorenzo, S.; Sindona, A.; Paganelli, S.; Giorgi, G.L.; Plastina, F. Many-qubit quantum state transfer via spin chains. *Phys. Scr.* **2015**, *T165*, 014036. [CrossRef]

14.   Lorenzo, S.; Apollaro, T.J.G.; Trombettoni, A.; Paganelli, S. 2-qubit quantum state transfer in spin chains and cold atoms with weak links. *Int. J. Quantum Inf.* **2017**, *15*, 1750037. [CrossRef]

15.   Lorenzo, S.; Apollaro, T.; Paganelli, S.; Palma, G.; Plastina, F. Transfer of arbitrary two qubit states via a spin chain. *Phys. Rev. A* **2015**, *91*, 42321. [CrossRef]

16.   Sousa, R.; Omar, Y. Pretty good state transfer of entangled states through quantum spin chains. *New J. Phys.* **2014**, *16*, 123003. [CrossRef]

17.   Vieira, R.; Rigolin, G. Robust and efficient transport of two-qubit entanglement via disordered spin chains. *Quantum Inf. Process.* **2019**, *123*. [CrossRef]

18.   Vieira, R.; Rigolin, G. Almost perfect transport of an entangled two-qubit state through a spin chain. *Phys. Lett. A* **2018**, *382*, 2586–2594. [CrossRef]

19.   Vieira, R.; Rigolin, G. Almost perfect transmission of multipartite entanglement through disordered and noisy spin chains. *Phys. Lett. A* **2020**, *384*, 126536. [CrossRef]

20.   Apollaro, T.J.; Sanavio, C.; Chetcuti, W.J.; Lorenzo, S. Multipartite entanglement transfer in spin chains. *Phys. Lett. A* **2020**, *384*, 126306. [CrossRef]

21.   Almeida, G.M.; Souza, A.M.; de Moura, F.A.; Lyra, M.L. Robust entanglement transfer through a disordered qubit ladder. *Phys. Lett. A* **2019**, *383*, 125847. [CrossRef]

22.   Verma, H.; Chotorlishvili, L.; Berakdar, J.; Mishra, S.K. Qubit(s) transfer in helical spin chains. *Epl* **2017**, *119*. [CrossRef]

23.   Kay, A. Basics of perfect communication through quantum networks. *Phys. Rev. A* **2011**, *84*, 022337. [CrossRef]

24.   Lieb, E.; Schultz, T.; Mattis, D. Two soluble models of an antiferromagnetic chain. *Ann. Phys.* **1961**, *16*, 407–466. [CrossRef]

25.   Chetcuti, W.J.; Sanavio, C.; Lorenzo, S.; Apollaro, T.J.G. Perturbative many-body transfer. *New J. Phys.* **2020**, *22*, 033030. [CrossRef]

26.   Bruus, H.; Flensberg, K. *Many-Body Quantum Theory in Condensed Matter Physics—An Introduction*; World Scientific: Singapore, 2004.

27.   Apollaro, T.J.G.; Almeida, G.M.A.; Lorenzo, S.; Ferraro, A.; Paganelli, S. Spin chains for two-qubit teleportation. *Phys. Rev. A* **2019**, *100*, 052308. [CrossRef]

28.   Banchi, L.; Vaia, R. Spectral problem for quasi-uniform nearest-neighbor chains. *J. Math. Phys.* **2013**, *54*, 043501. [CrossRef]

29.   Weinberg, P.; Bukov, M. QuSpin: A Python package for dynamics and exact diagonalisation of quantum many body systems. Part II: bosons, fermions and higher spins. *Scipost Phys.* **2019**, *7*. [CrossRef]

30.   Zwick, A.; Álvarez, G.A.; Stolze, J.; Osenda, O. Spin chains for robust state transfer: Modified boundary couplings versus completely engineered chains. *Phys. Rev. A* **2012**, *85*, 012318. [CrossRef]

31.   Zwick, A.; Álvarez, G.A.; Bensky, G.; Kurizki, G. Optimized dynamical control of state transfer through noisy spin chains. *New J. Phys.* **2014**, *16*, 065021. [CrossRef]

32.   Almeida, G.M.; de Moura, F.A.; Lyra, M.L. Quantum-state transfer through long-range correlated disordered channels. *Phys. Lett. A* **2018**, *382*, 1335–1340. [CrossRef]

33.   Pavlis, A.K.; Nikolopoulos, G.M.; Lambropoulos, P. Evaluation of the performance of two state-transfer Hamiltonians in the presence of static disorder. *Quantum Inf. Process.* **2016**, *15*, 2553–2568. [CrossRef]

34.   Lorenzo, S.; Apollaro, T.J.G.; Sindona, A.; Plastina, F. Quantum-state transfer via resonant tunneling through local-field-induced barriers. *Phys. Rev. A* **2013**, *87*, 042313. [CrossRef]

35.   Acosta Coden, D.S.; Gómez, S.S.; Ferrón, A.; Osenda, O. Controlled quantum state transfer in XX spin chains at the Quantum Speed Limit. *Phys. Lett. Sect. A Gen. At. Solid State Phys.* **2021**, *387*, 127009. [CrossRef]

36.   Bayat, A.; Omar, Y. Measurement-assisted quantum communication in spin channels with dephasing. *New J. Phys.* **2015**, *17*, 103041. [CrossRef]

37.   Kay, A. Perfect coding for dephased quantum state transfer. *Phys. Rev. A* **2018**, *97*, 1–8. [CrossRef]

Article

Entropy Dynamics of Phonon Quantum States Generated by Optical Excitation of a Two-Level System

Thilo Hahn, Daniel Wigger and Tilmann Kuhn *

Institut für Festköpertheorie, Universität Münster, Wilhelm-Klemm-Str. 10 48149 Münster, Germany; t.hahn@wwu.de (T.H.); d.wigger@wwu.de (D.W.)
* Correspondence: tilmann.kuhn@uni-muenster.de

Received: 04 February 2020; Accepted: 27 February 2020; Published: 29 February 2020

Abstract: In quantum physics, two prototypical model systems stand out due to their wide range of applications. These are the two-level system (TLS) and the harmonic oscillator. The former is often an ideal model for confined charge or spin systems and the latter for lattice vibrations, i.e., phonons. Here, we couple these two systems, which leads to numerous fascinating physical phenomena. Practically, we consider different optical excitations and decay scenarios of a TLS, focusing on the generated dynamics of a single phonon mode that couples to the TLS. Special emphasis is placed on the entropy of the different parts of the system, predominantly the phonons. While, without any decay, the entire system is always in a pure state, resulting in a vanishing entropy, the complex interplay between the single parts results in non-vanishing respective entanglement entropies and non-trivial dynamics of them. Taking a decay of the TLS into account leads to a non-vanishing entropy of the full system and additional aspects in its dynamics. We demonstrate that all aspects of the entropy's behavior can be traced back to the purity of the states and are illustrated by phonon Wigner functions in phase space.

Keywords: phonons; two-level system; entropy; Wigner functions; entanglement

---

1. Introduction

Entropy is one of the most fundamental concepts in physics. According to the second law of thermodynamics, in a closed system, it never decreases, which has far reaching consequences, from the limited efficiency of thermodynamic machines [1] to cosmological implications [2,3]. Under thermal equilibrium conditions, it determines the state of a thermodynamic system: In a closed system, the realized state is the one with maximal entropy; in a system in thermal contact with a heat bath, the realized state results from an interplay between energy and entropy and is governed by the minimum of the free energy [4]. Under nonequilibrium conditions, the second law of thermodynamics prohibits the decrease of the entropy of a closed system; however, this does not hold for the entropy of a subsystem which is interacting with other subsystems or with its surroundings [5,6]. In this case, the study of the dynamics of the entropy of these subsystems provides valuable information on the evolution of the nature of the system's state [7].

From the point of view of information science, entropy is closely related to the imperfect knowledge about a system [8,9]. As such, it plays a key role in all fields related to information processing and communication, and, in particular, in the highly topical fields of quantum information and communication, where the entropy is closely related to phenomena like purity of quantum states, entanglement, and decoherence [10].

In this paper, we study the entropy dynamics in a prototypical model of quantum mechanics and quantum information theory. It consists of two subsystems, a quantum-mechanical two-level system (i.e., a representation of a qubit) which can be manipulated by an external field (e.g., a light field) and which is coupled to a harmonic oscillator (e.g., a single phonon mode or a nanomechanical oscillator). The generation of specific quantum states of such a harmonic oscillator and the manipulation of these states has recently attracted much interest [11–14]. Prominent examples are coherent states and Schrödinger cat states, i.e., superpositions of coherent states. We analyze the entropy dynamics of the two subsystems after excitation with a short optical pulse or a pair of such pulses. In particular, we compare the case of a unitary evolution in the absence of damping processes, when the coupled system remains in a pure state, with the case of a decaying two-level system resulting in a mixed state also of the combined system. We will show that the analysis of the time-dependent entropy provides interesting insight into the nature of the quantum state of the two subsystems.

2. Theory

We consider a two-level system (TLS) which can be excited and de-excited by a resonant optical field **E**. Additionally, a pure dephasing coupling to a single phonon (ph) mode is taken into account. Thus, the Hamiltonian reads [15]

$$H = \hbar\Omega \left|x\right\rangle \left\langle x\right| - \left[\mathbf{M} \cdot \mathbf{E}(t) \left|x\right\rangle \left\langle g\right| + \mathbf{M}^* \cdot \mathbf{E}^*(t) \left|g\right\rangle \left\langle x\right| \right] + \hbar\omega_{\mathrm{ph}} \hat{b}^\dagger \hat{b} + \hbar g \left(\hat{b} + \hat{b}^\dagger\right) \left|x\right\rangle \left\langle x\right| . \tag{1}$$

The states $\left|g\right\rangle$ and $\left|x\right\rangle$ describe ground and excited state of the TLS with an energy splitting of $\hbar\Omega$, respectively. The time dependent optical driving is mediated by the dipole matrix element **M**. Phonons with the discrete energy $\hbar\omega_{\mathrm{ph}}$ are created and annihilated by $\hat{b}^\dagger$ and $\hat{b}$, respectively. For simplicity, the coupling constant of the exciton-phonon interaction g is supposed to be real.

Such a TLS system coupled to a single bosonic mode is a prototypical model that can be considered for the description of various solid state systems. For the TLS, one might think of an exciton in a single semiconductor quantum dot [16] or excitations of defects in insulators, like diamond [17] or hexagonal boron nitride [18], while the phonon could be an optical mode [19], a local mode [20], a van Hove singularity [21], or the mechanical excitation of a microresonator [22].

Phonon-induced transitions between the states $\left|g\right\rangle$ and $\left|x\right\rangle$ of the TLS are negligible because of the strong energy mismatch between the exciton energy, which is of the order of one or a few electronvolts (eV), while phonon energies range from a few micro-electronvolts (µeV) (for micromechanical resonators) up to a few tens of milli-electronvolts (meV) (for optical phonon modes). The linear coupling in the phonon displacement reflects typical electron-phonon interaction mechanisms in solids like deformation potential coupling, piezoelectric coupling, or Fröhlich coupling [23]. Although extensions of this model have been considered that take a quadratic coupling to the phonons into account [24–27], the original independent boson model [28] in Equation (1) is successfully used in different contexts. It reproduces recent linear and nonlinear spectroscopy signals [18,29], Rabi oscillations [30], and rotations [31] in excellent agreement with experiments, to name just a few.

The possible states of the entire system can be separated into the phonons forming product states with the ground state of the TLS and those forming product states with the excited state

$$\left|g\right\rangle \otimes \left|ph_g\right\rangle \qquad \text{and} \qquad \left|x\right\rangle \otimes \left|ph_x\right\rangle . \tag{2}$$

From the full density matrix of the system ρ, we can calculate the one of a subsystem by tracing over the respective other, i.e.,

$$\rho_{\mathrm{TLS}} = \mathrm{Tr}_{\mathrm{ph}}(\rho) \qquad \text{and} \qquad \rho_{\mathrm{ph}} = \mathrm{Tr}_{\mathrm{TLS}}(\rho) . \tag{3}$$

In the same way as in Reference [32], we model a decay of the excited state with the rate Γ via the Lindblad dissipator

$$\mathcal{D}(\rho) = \Gamma \left[\left|g\right\rangle \left\langle x\right| \rho \left|x\right\rangle \left\langle g\right| - \frac{1}{2}\left\{ \left|x\right\rangle \left\langle x\right|, \rho \right\} \right], \tag{4}$$

leading to the master equation for the density matrix:

$$\frac{\mathrm{d}}{\mathrm{d}t}\rho = \frac{1}{i\hbar}[H,\rho] + \mathcal{D}(\rho) \, . \tag{5}$$

We will not take an additional phenomenological pure dephasing of the TLS into account. We will study different regimes of decay rates Γ compared to the characteristic phonon frequency ω_{ph}. On the one hand, when describing, for example, optical phonons with energies in the range of tens of meV the decay time of the TLS is typically much longer than a phonon period, i.e., $\Gamma \ll \omega_{\mathrm{ph}}$ [18]. On the other hand, when considering typical mechanical resonators with phonon energies in the μeV range we have $\Gamma > \omega_{\mathrm{ph}}$ [33].

As we have explained in Reference [32], the entire quantum state and especially the Wigner function of the phonons can be calculated analytically when considering a series of ultrafast laser pulses to drive the TLS. Especially if the pulse duration is much shorter than the phonon period, the pulses can be approximated by delta-functions as

$$\frac{\mathbf{M} \cdot \mathbf{E}(t)}{\hbar} = \sum_j \frac{\theta_j}{2} \exp\left[-i\left(\Omega - \frac{g^2}{\omega_{\mathrm{ph}}}\right)t + i\phi_j\right]\delta(t - t_j) \, . \tag{6}$$

The pulses excite the TLS at times t_j with pulse areas θ_j and phases ϕ_j. By this choice and the introduction of the generating functions

$$Y_\alpha(t) = \left\langle |g\rangle\langle x| \exp(-\alpha^* \hat{b}^\dagger)\exp(\alpha \hat{b})\right\rangle \, , \tag{7a}$$

$$C_\alpha(t) = \left\langle |x\rangle\langle x| \exp(-\alpha^* \hat{b}^\dagger)\exp(\alpha \hat{b})\right\rangle \, , \tag{7b}$$

$$F_\alpha(t) = \left\langle \exp(-\alpha^* \hat{b}^\dagger)\exp(\alpha \hat{b})\right\rangle \, , \tag{7c}$$

with $\langle \hat{A}\rangle = \mathrm{Tr}\left(\rho\hat{A}\right)$ denoting the expectation value of an operator $\hat{A}$, a closed system of partial differential equations for the time-evolution of the generating functions is obtained and all phonon assisted density matrices can be calculated analytically without approximations. Note that F_α contains the entire information on the phonon system, while C_α describes the phonon assisted occupation of the excited state, i.e., the phonons in $|x\rangle \otimes |ph_x\rangle$, and Y_α is the phonon assisted coherence.

Our analysis of the phonon quantum states is based on their Wigner function [34]

$$W(U, \Pi) = \frac{1}{4\pi}\int\limits_{-\infty}^{\infty} \left\langle U + \frac{X}{2}\middle|\rho_{\mathrm{ph}}\middle|U - \frac{X}{2}\right\rangle \exp\left(-\frac{i}{2}X\Pi\right)\mathrm{d}X \, , \tag{8}$$

which is a quasi-probability distribution in the phase space defined by the quadratures $\hat{u}$ and $\hat{\pi}$ and their respective eigenstates

$$\hat{u} = \hat{b} + \hat{b}^\dagger \, , \qquad\qquad \hat{\pi} = \frac{1}{i}(\hat{b} - \hat{b}^\dagger) \, , \tag{9a}$$

$$\hat{u}|U\rangle = U|U\rangle \, , \qquad\qquad \hat{\pi}|\Pi\rangle = \Pi|\Pi\rangle \, . \tag{9b}$$

Due to their definition by the phonon annihilation and creation operators, the quantities U and Π directly correspond to the lattice displacement and momentum, respectively. The generating function F_α, at the same time, is a characteristic function of the Husimi Q function [35]. From this, we can directly calculate the instructive Wigner distribution analytically for a given pulse sequence via [34]:

$$W(U, \Pi, t) = \frac{1}{4\pi^2}\int\limits_{-\infty}^{\infty}\!\!\int \exp\left(-\frac{|\alpha|^2}{2}\right)F_\alpha(t)\exp\left\{i\left[\mathrm{Re}(\alpha)\Pi + \mathrm{Im}(\alpha)U\right]\right\}\mathrm{d}^2\alpha \, . \tag{10}$$

In the same way, we can isolate the Wigner function W_x for the phonons associated with the TLS being in the excited state $|x\rangle$ by choosing C_α instead of F_α in Equation (10). By doing the same, but choosing Y_α, we define W_p as the Wigner function of the phonon assisted coherence. In summary, we have

$$F_\alpha \to W \,, \tag{11a}$$
$$C_\alpha \to W_\mathrm{x} \,, \tag{11b}$$
$$Y_\alpha \to W_\mathrm{p} \,, \tag{11c}$$
$$F_\alpha - C_\alpha \to W - W_\mathrm{x} = W_\mathrm{g} \,, \tag{11d}$$

where W_g is the Wigner function of the phonons associated with the TLS being in the ground state $|g\rangle \otimes |ph_\mathrm{g}\rangle$.

Following the original definition by von Neumann [36], we investigate the time-dependent entropy of our coupled quantum system defined by

$$S = -\mathrm{Tr}\big[\rho \ln(\rho)\big] \,. \tag{12}$$

In general, subadditivity states that the entropy of the full system is a lower boundary for the sum of the entropies of the subsystems [37]:

$$S(\rho) \leq S(\rho_\mathrm{TLS} \otimes \rho_\mathrm{ph}) = S(\rho_\mathrm{TLS}) + S(\rho_\mathrm{ph}) \,. \tag{13}$$

It is important to note that for every pure quantum state ρ_pure, the entropy vanishes, i.e.,

$$S_\mathrm{pure} = -\mathrm{Tr}\big[\rho_\mathrm{pure} \ln(\rho_\mathrm{pure})\big] = 0 \,, \tag{14}$$

and the entropies of the subsystems coincide if the state of the full system is pure [37]

$$S^\mathrm{ph} = S(\rho_\mathrm{ph}) = S(\rho_\mathrm{TLS}) \,. \tag{15}$$

To show this, following Reference [37] for an arbitrary composed system, we decompose the pure state of the entire system $|\psi\rangle$ into an orthonormal basis $|\psi\rangle = \sum_{i,k} C_{ik} |\phi_i\rangle |\chi_k\rangle$ with the coefficient matrix $C = (C_{ik})$. Here, $|\phi_i\rangle$ and $|\chi_k\rangle$ are basis states of the two subsystems, respectively. From the complete density matrix $\rho = (\rho_{ikjl})$, one obtains a reduced density matrix by tracing over the respective other subsystem $\rho_1 = Tr_2(\rho)$:

$$\rho_{ikjl} = \langle\phi_i|\langle\chi_k|\rho|\chi_l\rangle|\phi_j\rangle = C_{ik}C_{jl}^* \,, \tag{16a}$$
$$\Rightarrow \rho_{1,ij} = \langle\phi_i|\rho_1|\phi_j\rangle = \sum_k C_{ik}C_{jk}^* = (CC^\dagger)_{ij} \,. \tag{16b}$$

Analogously the density matrix of the other subsystem is $\rho_2 = C^\dagger C$. Any non-vanishing eigenvalue λ of ρ_1 with the eigenvector $\mathbf{y}$ is then also an eigenvalue of ρ_2 with eigenvector $\mathbf{z} = C^\dagger \mathbf{y}$ because

$$CC^\dagger \mathbf{y} = \lambda \mathbf{y} \,, \tag{17a}$$
$$\Rightarrow C^\dagger C \mathbf{z} = C^\dagger C (C^\dagger \mathbf{y}) = C^\dagger (CC^\dagger)\mathbf{y} = \lambda C^\dagger \mathbf{y} = \lambda \mathbf{z} \,, \tag{17b}$$

and vice versa.

In our particular system, we can choose the TLS's states as the $|\phi_i\rangle \in \{|g\rangle, |x\rangle\}$ and a Fock basis for the phonon system $|\chi_k\rangle \in \{|0\rangle, |1\rangle, \ldots\}$. This means that the coefficient matrix C consists of $2 \times \infty$ elements. The entropy of the TLS, i.e., of its density matrix:

$$CC^\dagger = \rho_{TLS} = \begin{pmatrix} 1-c & p \\ p^* & c \end{pmatrix}, \qquad \text{with} \quad c = \langle |x\rangle\langle x| \rangle \,, \; p = \langle |g\rangle\langle x| \rangle \,, \tag{18a}$$

can be easily calculated via [37]

$$S(\rho_{TLS}) = -\lambda_+ \ln(\lambda_+) - \lambda_- \ln(\lambda_-) \,, \tag{18b}$$

where

$$\lambda_\pm = \frac{1}{2} \pm \sqrt{\frac{1}{4} + c^2 - c + |p|^2} = \frac{1}{2} \pm \frac{1}{2}|\mathbf{v}| \tag{18c}$$

are the two eigenvalues of the TLS's density matrix with the Bloch vector $\mathbf{v} = (2\mathrm{Re}(p), 2\mathrm{Im}(p), 2c - 1)$. For a pure state of the full system, according to Equation (17), the phonon density matrix has, despite being a quadratic infinite dimensional matrix, only two non-vanishing eigenvalues $\lambda_\pm$. So, as long as the entire system is in a pure state and we know the entropy of the TLS via Equation (18b), we can derive the entropy of the phonon system by Equation (15).

For an arbitrary, non-pure state of the entire system, the calculation of the entropy in a system with infinite dimensions is far from being trivial. This is the case if, already, the initial state is a statistical mixture, e.g., at non-vanishing temperature or when dephasing leads to a statistical mixture. Therefore, approximations have been discussed and a reasonable version is given by the linear entropy [37]:

$$S_{\mathrm{lin}} = \mathrm{Tr}(\rho) - \mathrm{Tr}(\rho^2) = 1 - \mathrm{Tr}(\rho^2) = 1 - \langle \rho \rangle \,. \tag{19}$$

To get an impression of this approximation, Figure 1a shows the function of the full entropy $-\xi\ln(\xi)$ and the one for the linear entropy $\xi - \xi^2$; similar to the presentation in Reference [37], note that $\mathrm{Tr}(\rho) = 1$. We find that the function of the approximated linear entropy in red is always smaller than the full entropy in blue. Therefore, we expect that the linear entropy under-estimates the full entropy.

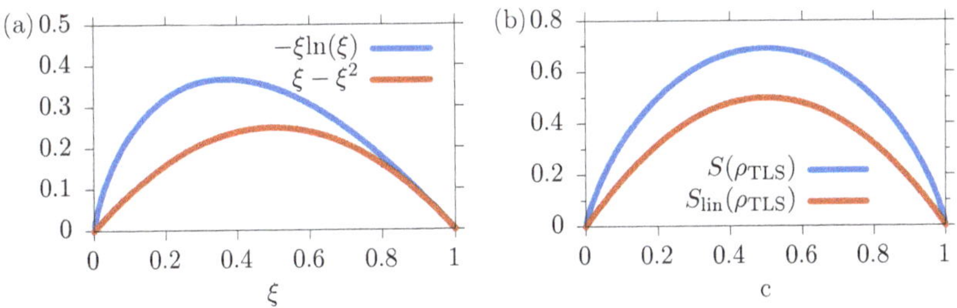

Figure 1. (a) Functions for the full entropy in blue and the linear entropy in red. (b) Entropies of the two-level system (TLS) with $p = 0$, full entropy in blue and linear entropy in red.

We can directly compare the linear and the full entropy of the isolated TLS from Equation (18b), as shown in Figure 1b. There, the full entropy is plotted in blue and the linear one in red as functions of the excited state occupation c, both for $p = 0$. In the limiting cases of full inversion $c = 1$ and no inversion $c = 0$, the state is pure and both entropies vanish. For all other occupations, the TLS is in a statistical mixture, and the entropy is non-zero, and it is $S_{\mathrm{lin}} \leqslant S$. In the case of an equally distributed mixture, i.e., $c = 0.5$, the entropies are maximal and reach values of $S_{\mathrm{lin}} = 0.5$ and $S = \ln(2) \approx 0.7$.

The biggest advantage of this linear entropy for our study is that it can be directly calculated from the phonons' Wigner functions due to the trace-product rule [38] via

$$S_{\mathrm{lin}}^{\mathrm{ph}} = 1 - 4\pi \iint\limits_{-\infty}^{\infty} W(U, \Pi)^2 \, \mathrm{d}U \, \mathrm{d}\Pi \,. \tag{20}$$

Note that the prefactor 4π depends on the definition of the quadratures U and Π, i.e., the scaling of the phase space.

With the separation into TLS and phonon system, the linear entropy is calculated via

$$S_{\mathrm{lin}} = 1 - \mathrm{Tr}_{\mathrm{ph}}\left[\mathrm{Tr}_{\mathrm{TLS}}(\rho^2)\right] . \tag{21}$$

For this, we again consider the density matrix of the full system as

$$\rho = (1-c)\,|g\rangle\langle g| \otimes \rho_{\mathrm{ph,g}} + c\,|x\rangle\langle x| \otimes \rho_{\mathrm{ph,x}} + p^*\,|g\rangle\langle x| \otimes \rho_{\mathrm{ph,p}} + p\,|x\rangle\langle g| \otimes \rho^\dagger_{\mathrm{ph,p}} , \tag{22a}$$

$$\Rightarrow \mathrm{Tr}_{\mathrm{TLS}}(\rho^2) = (1-c)^2\rho^2_{\mathrm{ph,g}} + c^2\rho^2_{\mathrm{ph,x}} + |p|^2(\rho_{\mathrm{ph,p}}\rho^\dagger_{\mathrm{ph,p}} + \rho^\dagger_{\mathrm{ph,p}}\rho_{\mathrm{ph,p}}) , \tag{22b}$$

leading to

$$\begin{aligned}
\mathrm{Tr}_{\mathrm{TLS}}(\rho) - \mathrm{Tr}_{\mathrm{TLS}}(\rho^2) &= (1-c)\rho_{\mathrm{ph,g}} - (1-c)^2\rho_{\mathrm{ph,g}} \\
&+ c\rho_{\mathrm{ph,x}} - c^2\rho^2_{\mathrm{ph,x}} \\
&- |p|^2(\rho_{\mathrm{ph,p}}\rho^\dagger_{\mathrm{ph,p}} + \rho^\dagger_{\mathrm{ph,p}}\rho_{\mathrm{ph,p}}) .
\end{aligned} \tag{22c}$$

We can now use the separate parts of the Wigner function from Equation (11) to define entropies

$$S^{\mathrm{i}}_{\mathrm{lin}} = \int\!\!\!\int_{-\infty}^{\infty} \left[W_{\mathrm{i}}(U,\Pi) - 4\pi W_{\mathrm{i}}(U,\Pi)^2\right] dU\,d\Pi , \qquad \text{with } \mathrm{i} \in \{g,x\} , \tag{22d}$$

$$S^{\mathrm{p}}_{\mathrm{lin}} = -4\pi \int\!\!\!\int_{-\infty}^{\infty} |W_{\mathrm{p}}(U,\Pi)|^2\,dU\,d\Pi . \tag{22e}$$

Note that the polarization Wigner function W_{p} is a complex quantity. With this and the definitions of the generating functions in Equation (7), we can write the linear entropy in Equation (21) as

$$S_{\mathrm{lin}} = S^{g}_{\mathrm{lin}} + S^{x}_{\mathrm{lin}} + 2S^{\mathrm{p}}_{\mathrm{lin}} . \tag{23}$$

To briefly summarize, for pure states of the entire system, i.e., without dephasing or decay of the TLS, we can calculate the full entropy of the phonon state S_{ph} via Equation (15). If the state is not pure, we can at least calculate the linear entropy from the Wigner functions. We can further distinguish between the linear entropy of the full system S_{lin} in Equation (23) and the one of the phonons $S^{\mathrm{ph}}_{\mathrm{lin}}$ in Equation (20).

3. Results and Discussion

3.1. Single Pulse Excitation

We start our study with the most basic situation, where the TLS is excited by a single optical pulse. It is well known from previous works that a single ultrafast excitation in general creates a statistical mixture of coherent states in the phonon system. The excitation of the TLS means for the phonons a shift of the equilibrium position determined by the dimensionless coupling strength $\gamma = g/\omega_{\mathrm{ph}}$. If not stated differently, in the following, we fix this value to $\gamma = 2$ in order to separate the different parts of the Wigner function in phase space, as will be seen later. Although $\gamma = 2$ is a rather large value for quantum dots and optical phonons, the general physics explained in this paper will not depend on this value. Some effects might be strengthened or weakened with a different choice of the coupling strength, as will be highlighted later.

3.1.1. Phonons Generated by a Non-Decaying TLS

In the first step, we neglect the decay of the excited state by choosing $\Gamma = 0$. In this situation, the state of the full system, including the TLS and the phonons, is pure. Therefore, the full and the linearized entropy are zero and Equation (15) holds, meaning that the entropy of the TLS and that of the phonons is the same. Figure 2 recapitulates the phonon dynamics for a pulse area of $\theta = \pi/2$, i.e., an inversion of the TLS of 50% or $c = 0.5$, from Reference [13]. The phonon's Wigner function reads:

$$W(U, \Pi, t) = \frac{1}{4\pi} \left\{ \exp\left[-\frac{1}{2}(U^2 + \Pi^2) \right] \right.$$
$$\left. + \exp\left(-\frac{1}{2}\{U - 2\gamma[1 - \cos(\omega_{\text{ph}}t)]\}^2 - \frac{1}{2}[\Pi - 2\gamma\sin(\omega_{\text{ph}}t)]^2 \right) \right\}, \tag{24}$$

and its dynamics are shown at five different times in Figure 2a. Before the optical excitation, the phonons are in the vacuum state represented by the Gaussian Wigner function in the center of the phase space. Half of the weight of the phonon's Wigner distribution is brought into the excited state subspace by the optical pulse. This makes them move as a coherent state around the new equilibrium position, which is shifted by 2γ in U-direction. This trajectory is marked as black circle in the figure. The other half of the phonon state remains associated with the ground state of the TLS and, therefore, stays in the vacuum state. The full phonon state after tracing over the TLS states is a statistical mixture of the vacuum state and a coherent state moving around the shifted equilibrium position. After a full phonon period at $t = t_{\text{ph}}$, the Wigner function agrees with the initial situation because the coherent state moves through the origin and overlaps with the vacuum state. The phonon's influence on the properties of the TLS is shown in Figure 2b. While the occupation of the excited state stays constant at $c = 0.5$, the polarization $|p|$ starts at 0.5 directly after the optical excitation at $t = 0$ and drops rapidly to almost zero in the following. This dephasing is inverted towards $t = t_{\text{ph}}$, resulting in a full rephasing to $|p| = 0.5$. While the coherent states separate in phase space, coherence is lost from the TLS, which already shows that the overlap of the different parts of the Wigner function plays an important role for the properties of the entire system.

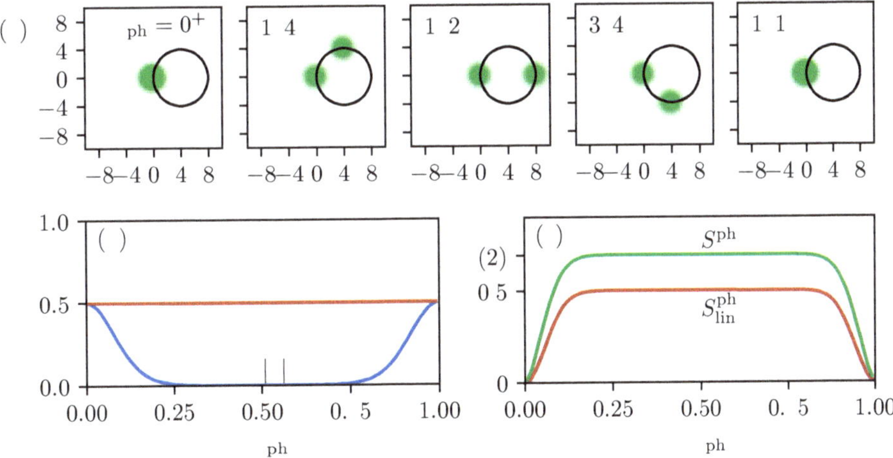

Figure 2. (a) Dynamics of the phonon Wigner function after a single pulse excitation of the TLS. (b) TLS dynamics with the excited state occupation c in red and the polarization $|p|$ in blue. (c) Entropies of the phonon system, S^{ph} in green and $S^{\text{ph}}_{\text{lin}}$ in red.

Finally, in Figure 2c, we show the entropy of the phonons which, as mentioned above, agrees with the entropy of the TLS. As the initial phonon state is pure, both entropies, S^{ph} in green and S^{ph}_{lin} in red, start at zero at $t = 0$. While the different parts in phase space separate, the entropy grows to $S^{ph}_{lin} \approx 0.5$ because the phonons are in a statistical mixture that must have a non-vanishing entropy. The full entropy follows the same dynamics as the linear one but is always larger, as previously explained, and grows to $S^{ph} \approx \ln(2) \approx 0.7$. Reaching $t = t_{ph}$, the entropies drop to zero again. The reason is the recovered overlap of the two parts of the Wigner function. Finally, at $t = t_{ph}$, the phase space representation cannot be distinguished from the vacuum state. Therefore, the entropy also has to agree with the one of the pure vacuum state being zero.

3.1.2. Phonons Generated by a Decaying TLS

In the previous section, without any decay or additional pure dephasing of the TLS, the quantum state of the entire system remained pure, resulting in a vanishing entropy of the full system. It also allowed us to easily calculate the full entropies for TLS and phonons. In this section, we consider a non-vanishing decay rate of the occupation of the excited state into the ground state, which naturally results in a statistical mixture in the TLS's quantum state that also imprints onto the phonons. Therefore, for the phonons, we can only calculate linear entropies, according to Equation (20). In Reference [32], we explained how the Wigner function evolves during the decay process in the TLS. Therefore, we consider the same optical excitation with a pulse area of $\theta = \pi$, which initially fully inverts the TLS. Without any decay, the Wigner function would read

$$W(U, \Pi, t) = \frac{1}{2\pi} \exp\left(-\frac{1}{2}\{U - 2\gamma[1 - \cos(\omega_{ph}t)]\}^2 - \frac{1}{2}[\Pi - 2\gamma\sin(\omega_{ph}t)]^2\right) , \qquad (25)$$

being a single Gaussian moving on a circle around the shifted equilibrium position of the excited state. In Figure 3a, Wigner functions for different decay rates Γ are shown at $t = 10t_{ph}$. We find that the phonon state gets smeared out in phase space. For rapid decays on the left, the phonons almost completely stay in the vacuum state and look more or less like a coherent Gaussian distribution. When looking at the corresponding linear entropy dynamics in Figure 3b, in bright red, we see that it only increases slightly after the optical excitation at $t = 0$. When slowing down the decay process, i.e., moving in Figure 3a more to the right, the Wigner function smears out more and more. Accordingly it looks less and less like a coherent state which also leads to increasing entropies in (b) when going from bright to dark colors. Additionally, we find that the final entropy value is reached slower because it follows the decay of the TLS. Especially for the slowest considered decay of $\Gamma = 0.1\omega_{ph}$, where the full decay takes several phonon periods, the dynamics of the linear phonon entropy develop minima at full phonon periods $t = nt_{ph}$. These are the times when the Wigner function in Figure 3a starts overlapping itself. This is exemplarily shown in Figure 3c for $t = t_{ph}$, where the thick Gaussian part is the oscillating coherent state. This is the first time it intersects with the circular distribution that has already decayed into $|g\rangle$. In agreement with the findings in Figure 2, this leads to the temporary reduction of the entropy in Figure 3b.

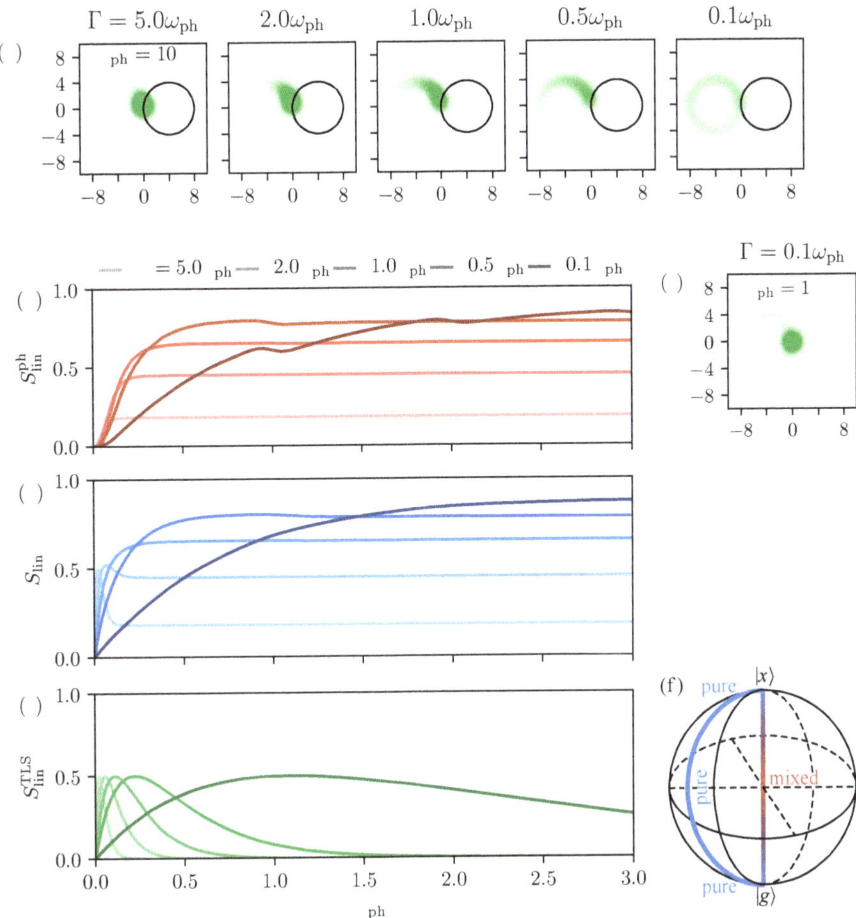

Figure 3. (**a**) Phonon Wigner function after a full decay of the TLS at $t = 10t_{\mathrm{ph}}$ for different decay rates Γ as given in the picture. (**b**) Linear phonon entropy as a function of time after the pulse. The decay rate decreases from bright to dark colors. (**c**) Exemplary Wigner function for a slow decay rate of $\Gamma = 0.1\omega_{\mathrm{ph}}$ at $t = t_{\mathrm{ph}}$. (**d**) Same as (**b**) but for the full linear entropy. (**e**) Same as (**b**) but for the linear entropy of the TLS. (**f**) Bloch sphere of the TLS to illustrate the purity of different states; blue shows pure and red mixed states.

Comparing the linear entropy of the phonon system in Figure 3b with the linear entropy of the entire system S_{lin} in (d), we basically find the same overall behavior. The dynamics start at zero, the final values increase for smaller decay constants, and the final values are reached later. However, both for small and large Γ, we find qualitative differences. Starting with small Γ in dark blue, especially for $\Gamma = 0.1\omega_{\mathrm{ph}}$, the curve constantly grows without developing any minima. This shows that the reduction of phonon entropy due to the overlapping Wigner functions does not attain to the full entropy. The reason for this is that the reduction of the phonon entropy due to overlapping parts of the Wigner function only happens because the state information of the TLS has been traced out. Taking the entire coupled system into account, the overlapping phonon parts belong to different states of the TLS and can therefore be told apart. Therefore, the phonons do not lead to a depression of the full linear entropy. Conversely, for large decay rates, in bright blue, especially $\Gamma = 5\omega_{\mathrm{ph}}$, we find that, on a short timescale around $t = 0.1t_{\mathrm{ph}}$, a pronounced maximum appears in S_{lin}. This effect is not found in the phonon part in Figure 3b and therefore stems from the TLS contribution. To understand this, in Figure 3e, we plot the linear entropy of the TLS as green lines. The bright and dark colors agree with the ones in Figure 3b,d. In addition, we consider the schematic Bloch vector representation of the TLS state from Equation (18c) in Figure 3f. The z direction of the Bloch sphere depicts the occupation of the states, where the south pole is a pure ground state $|g\rangle$ and the north pole a pure excited state $|x\rangle$. Between these points, all Bloch vectors that are on the surface of the sphere (blue line) are superpositions $|\chi\rangle = N(\alpha|g\rangle + \beta|x\rangle)$ and are therefore pure. In the other extreme case of the line directly connecting north and south pole (red), the system is in a statistical mixture of $|g\rangle$ and $|x\rangle$. In the center of the Bloch sphere, the TLS is in both states with equal probability, resulting in the lowest purity. For the entropy of the TLS in Figure 3e, this means that, directly after the excitation into the excited state and after the full decay, the entropy is zero. In between, the system evolved through a statistical mixture, which has a non-vanishing entropy. The linear entropy reaches maxima with $S_{\mathrm{lin}}^{\mathrm{TLS}} = 0.5$, in agreement with the result in Figure 1b.

If we want to determine the final linear entropy $S_{\mathrm{lin}}^{\mathrm{ph},\infty} = S_{\mathrm{lin}}^{\mathrm{ph}}(t \to \infty)$ of the phonon state after the TLS is fully decayed into the ground state, we can investigate the dynamics in phase space. Note that the final entropy is only carried by the phonon part because the TLS is in the pure ground state. As schematically shown in Figure 4a, the movement of the coherent state on the circle in the excited state subspace and the accompanied decay into the ground state leads to a distribution that can be seen as a continuous distribution of coherent states with decreasing amplitude. We can parametrize the circular motion of the Wigner function including the decay of the amplitude by

$$W_\infty(U,\Pi) = \Gamma \int_0^\infty \exp(-\Gamma t) W_{(U_0(t),\Pi_0(t))}\, dt \, , \tag{26}$$

where $W_{(U_0(t),\Pi_0(t))}$ is a Gaussian centered around $(U,\Pi) = (U_0(t),\Pi_0(t))$. With the circular trajectory in Equation (25), we have to consider

$$U_0(t) = 2\gamma[1 - \cos(\omega_{\mathrm{ph}}t)] \qquad \text{and} \qquad \Pi_0(t) = 2\gamma \sin(\omega_{\mathrm{ph}}t)\,. \tag{27}$$

Note that, to retrieve the Wigner distribution in the ground state in Figure 3a, one has to mirror the schematic in Figure 4a. However, the final linear entropy remains unaffected because it only depends on the general shape of the distribution. With this, the final linear entropy reads

$$S_{\mathrm{lin}}^{\mathrm{ph},\infty} = 1 - 4\pi \iint\limits_{-\infty}^{\infty} W_\infty^2(U,\Pi)\, dU d\Pi$$

$$= 1 - 4\pi\Gamma^2 \iint\limits_{-\infty}^{\infty} \iint\limits_{0}^{\infty} \exp\left[-\Gamma(t+t')\right] W_{(U_0(t),\Pi_0(t))} W_{(U_0(t'),\Pi_0(t'))}\, dtdt'\, dU d\Pi\,. \tag{28}$$

The integral over U and Π describes the overlap of two coherent states in phase space. In general, two coherent states with a phase space distance of a have an overlap of [34]

$$4\pi \iint\limits_{-\infty}^{\infty} W_{(0,0)} W_{(a,0)}(U,\Pi)\, dU d\Pi = \exp\left[-\frac{a^2}{4}\right]\,. \tag{29}$$

Therefore, the entropy becomes

$$S_{\mathrm{lin}}^{\mathrm{ph},\infty} = 1 - \Gamma^2 \iint\limits_{0}^{\infty} \exp\left[-\Gamma(t+t')\right] \exp\left(-\left\{2\gamma\sin\left[\frac{1}{2}\omega_{\mathrm{ph}}(t-t')\right]\right\}^2\right)\, dtdt'\,. \tag{30}$$

The sine function in the exponent can be approximated by a linear function for small frequencies or short times. Note that, although the integration is carried out up to $t = \infty$, the exponential decay with Γ effectively limits the integrated time interval. Therefore, we expect this approximation to work well for sufficiently large Γ. For the motion of the Wigner function, this corresponds to a linear movement in phase space, as schematically shown in Figure 4b. The corresponding linear entropy can then be calculated to

$$S_{\mathrm{lin}}^{\mathrm{ph},\infty} \approx 1 - \Gamma^2 \int\limits_0^\infty\!\!\int \exp\left[-\Gamma(t+t')\right] \exp\left\{-[\gamma\omega_{\mathrm{ph}}(t-t')]^2\right\}\, dt\, dt'$$

$$= 1 - \frac{\sqrt{\pi}\,\Gamma}{2\gamma\omega_{\mathrm{ph}}} \exp\left[\left(\frac{\Gamma}{2\gamma\omega_{\mathrm{ph}}}\right)^2\right] \mathrm{erfc}\left(\frac{\Gamma}{2\gamma\omega_{\mathrm{ph}}}\right), \tag{31}$$

where $\mathrm{erfc}(x) = 1 - \mathrm{erf}(x)$ and $\mathrm{erf}(x)$ are the error function.

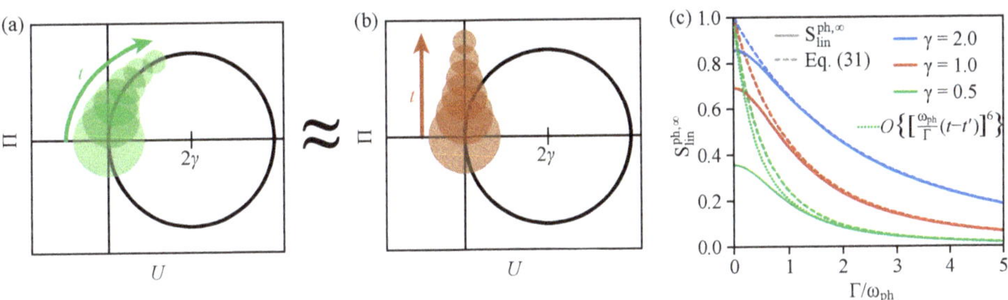

Figure 4. Final value of the entropy after the full decay of the TLS. (**a**) Schematic of the phase space dynamics of the Wigner function during the decay. (**b**) Approximated dynamics as a straight line. (**c**) Final linear entropy $S_{\mathrm{lin}}^{\mathrm{ph},\infty}$ as a function of the decay rate Γ, full simulation in solid and approximation from Equation (31) in dashed lines. Different coupling strengths are shown in blue, red, and green. The dotted green line shows the entropy for an expansion of the squared sine-function up to the sixth order.

The results for the final entropy $S_{\mathrm{lin}}^{\mathrm{ph},\infty}$ are shown in Figure 4c as a function of the decay rate Γ, where the full calculations according to the dynamics are shown as solid lines and the approximations from Equation (31) are the dashed lines. We show the three different coupling strengths $\gamma = 2$ (blue), $\gamma = 1$ (red), and $\gamma = 0.5$ (green). Comparing the different coupling strengths, we find that a stronger coupling leads to a larger final entropy because the Wigner function gets distributed over a larger area of phase space. As explained before, the approximation in Equation (31) works very good for large Γ, but we also see that the approximation works over a larger Γ range if the coupling strength is larger. The reason for this is that, for larger γ, the circle of the Wigner function's trajectory is larger, meaning that its curvature can be better approximated by a linear motion. For the smallest considered coupling strength $\gamma = 0.5$, in green, we additionally show the dotted line that stems from an approximation of the squared sine function up to the sixth order, which is the next non-divergent contribution in Equation (30). This curve is obviously a better approximation of the full calculation, in particular, in the range $1 \lesssim \Gamma/\omega_{\mathrm{ph}} \lesssim 2$. However, in agreement with all dashed lines, it reaches $S_{\mathrm{lin}}^{\mathrm{ph},\infty} = 1$ for $\Gamma = 0$, while all full linear entropies go to smaller values. This shows that any approximation of the sine function will not give accurate results for very small decay rates. In fact, it turns out that, for any finite (converging) order of the expansion of the squared sine function in Equation (30), $S_{\mathrm{ph}}^{\mathrm{lin},\infty} = 1$ is reached in the limit $\Gamma \to 0$. This can be understood by realizing that, for any finite order in the expansion of the trigonometric functions in Equation (27), the trajectory tends to infinity for $t \to \infty$, thus leading to a delocalized Wigner function. On the other hand, the correct

Wigner function in this limit is the doughnut-shaped function similar to the rightmost function in Figure 3a, for which the linear entropy can be calculated analytically, yielding

$$S_{\text{ph}}^{\text{lin},\infty} = 1 - e^{-2\gamma^2} I_0\!\left(2\gamma^2\right),$$

(32)

with the modified Bessel function of first kind and zeroth order I_0, in perfect agreement with the numerical results given in Figure 4c.

3.2. Two Pulse Excitation

The phonon quantum state gets more involved when a two-pulse excitation is considered. As extensively studied in Reference [13,32], an excitation with two pulses having pulse areas of $\theta_1 = \theta_2 = \pi/2$ and a delay of $t_2 - t_1 = t_{\text{ph}}/2$ leads to the generation of two Schrödinger cat states, each in one TLS subspace. A Schrödinger cat state is a coherent superposition of two coherent states, i.e., of the most classical states of a harmonic oscillator, and, as such, it is of high interest in all areas of quantum optics [39–41] and, more recently, also phononics [11].

3.2.1. Phonons Generated by a Non-Decaying TLS

We assume the same excitation scheme as just described and again disregard any decays of the TLS. The dynamics of the Wigner function are exemplarily shown in Figure 5a. Immediately before the second laser pulse reaches the TLS, the phonons are in the statistical mixture previously shown in Figure 2a. The second pulse creates a second coherent state in the excited state subspace of the TLS, but it also makes half of the coherent state in $|x\rangle$ go back to the ground state. Therefore, we end up with two coherent states in both subspaces, $|g\rangle$ and $|x\rangle$. As nicely seen in Figure 5a, the corresponding Wigner function shows two of the classic dumbbell structures of the cat state, two Gaussians and a striped structure of alternating positive (green) and negative (orange) values between them. These stripes indicate the interference between the two coherent states. The Wigner function in the ground state rotates around the origin, and the one in the excited state around the shifted equilibrium at $(U, \Pi) = (2\gamma, 0) = (4, 0)$ is marked by the black circle.

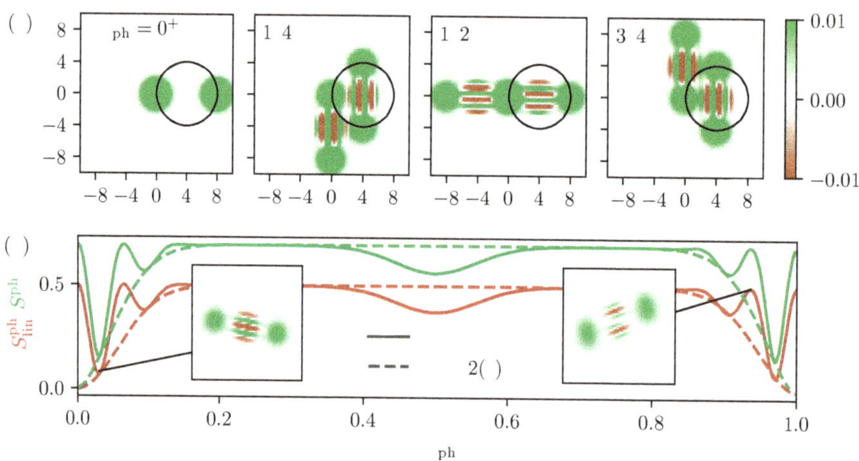

Figure 5. Two pulse excitation without decay. (**a**) Snapshots of the Wigner function after the second pulse. (**b**) Entropy dynamics after the second pulse, solid lines are the entropies of the cat states, dashed lines the statistical mixture from Figure 2c. The full phonon entropy is green and the linear one red.

The entropies of the phonon state are shown in Figure 5b as solid lines. The linear entropy in red starts at $S_{\mathrm{lin}} = 0.5$ in agreement with the entropy in Figure 2d at $t = t_{\mathrm{ph}}/2$ because at this time the second pulse excites the TLS. In contrast to the behavior after the first pulse, here, the entropy drops very rapidly and forms a sharp minimum after the second pulse. In total, the entropy performs two oscillations before remaining constant at the initial value. The same dynamics repeat themselves in an inverted form before reaching a full period. Another striking feature is a smaller depression around $t = t_{\mathrm{ph}}/2$. Overall, we find that, by the second pulse, the entropy of the phonon is temporarily reduced but never increased. The maximum entropy is here the one of a statistical mixture of two fully separated coherent states (see Figure 2), which is obviously the same for a statistical mixture of two cat states that are fully separated. This is the case during the times around $t/t_{\mathrm{ph}} = 1/4$ and $t/t_{\mathrm{ph}} = 3/4$ (see Figure 5a). Next, we discuss the reduced entropy around half a period in Figure 5b. Looking at the corresponding Wigner function in Figure 5a, we find that this is the time when one of the coherent states in the excited state system (moving on the circle) overlaps with the vacuum state (staying at the origin). This is the same effect as discussed in the previous sections, where the entropy shrank when the phonon states were overlapping in phase space. Finally, we have to understand the strong reductions of the entropy for times around full periods. To do so, we examine the two insets in Figure 5b that show snapshots of the Wigner function at the marked times, i.e., where the entropy is minimal and maximal. The left one at the minimum depicts a time where each of the two Gaussians starts to split into two, which cannot yet be resolved in the figure because their overlap is still too large. However, the interference terms between them also move apart. At $t = 0$, their negative and positive values are distributed in such a way that they exactly compensate each other (see $t = 0^{+}$ in Figure 5a). But, in the left inset, we see that at this time negative and positive values add up, respectively, making for an accurate alternating pattern. Comparing this structure of the Wigner function with one of the cat states in Figure 5a shows a strong resemblance. So, the reason for the strong decrease of the entropy is that, at these times, the Wigner function can only hardly be distinguished from a single Schrödinger cat state, which is a pure state. Likewise, we can analyze the Wigner function at a maximum of the entropy oscillation in the right inset. In addition, here, the Gaussians have a large overlap, but the stripes of the interferences are aligned in such a way that the line in the center has vanishing values. This strongly disagrees with the natural structure of a cat state interference and makes it easily distinguishable from that pure cat state. The full phonon entropy S^{ph} is shown as a green solid line. It follows the same dynamics as the linear one but is just scaled to larger values, as discussed before. Finally, let us remark on the additionally plotted dashed lines, which are the respective entropy curves from Figure 2c. They exactly form envelopes for the oscillations and therefore demonstrate that the oscillating dynamics are again a result of the separation process of the different Wigner functions in phase space.

The relative phase of the two laser pulses changes the phase in the cat states, i.e., the phase of the striped structure of the Wigner function. As long as the different parts of the phonon state are separated in phase space, the phase has no influence on the phonon entropy. The other crucial parameter of the phonon system is the coupling strength γ that determines the distance of the coherent states and the number of stripes in the interference term, as exemplarily shown in Figure 6a,b. The influence of an increased coupling strength is presented in Figure 6c, where the linear phonon entropy $S^{\mathrm{ph}}_{\mathrm{lin}}$ is plotted in the same way as in Figure 5b but only for times up to $t/t_{\mathrm{ph}} = 0.25$. The red curve shows $\gamma = 2$ from Figure 5b as a reference. Looking at the larger coupling in bright red ($\gamma = 4$) and a smaller coupling in dark red ($\gamma = 1.5$), we clearly see that the oscillation of the entropy gets faster when γ grows and more minima evolve. The reason is that the interference terms consist of more stripes that run through each other. At the same time, the envelope gets shorter for a larger γ, which can be traced back to the larger spread of the Wigner function in phase space. While the sizes of the interference terms and the Gaussians stay the same, as shown in Figure 6a,b, the radius of the trajectory increases. Because the angular frequency of the motion remains the same, the two interferences separate in a shorter time, and this time determines the envelope of the entropy in Figure 6c.

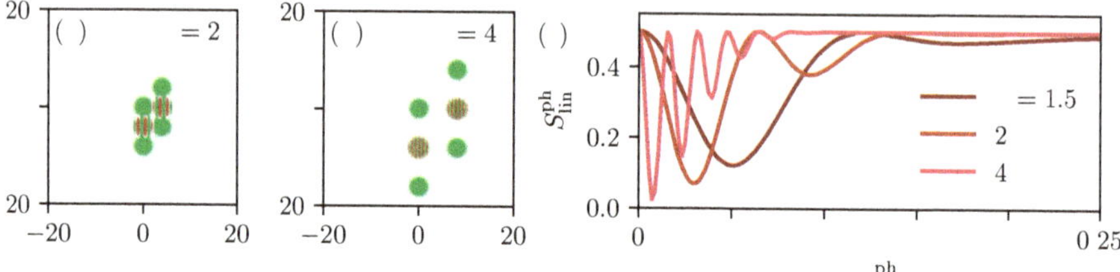

Figure 6. (**a**,**b**) Exemplary Wigner functions at $t = t_{\mathrm{ph}}/4$ for $\gamma = 2$ in (**a**) (same as in Figure 5a) and $\gamma = 4$ in (**b**). (**c**) Linear phonon entropy $S_{\mathrm{lin}}^{\mathrm{ph}}$ as a function of time after a two-pulse excitation as in Figure 5b. The coupling strength γ increases from dark to bright red.

3.3. Phonon Cat State Entropy Dynamics in a Decaying TLS

Next, we increase the complexity of the considered phonon state by analyzing the influence of the decay of the TLS on the entropy dynamics of cat states. For reasons of clarity, we consider a single Schrödinger cat state entirely in the excited state $|x\rangle \otimes |\mathrm{cat}\rangle$ as initial state without any optical excitation and account for a decay of the TLS into its ground state. Although this state cannot directly be prepared by optical pulses, in Reference [32] it is explained how it is constructed mathematically as initial state for the simulated decay dynamics. Some snapshots of the corresponding Wigner function dynamics are shown in Figure 7a for a small decay rate of $\Gamma = 0.1\omega_{\mathrm{ph}}$. As analyzed in Reference [32], the combined rotation and shift of the phonon equilibrium position due to the decay of the TLS finally leads to a Wigner function in the shape of an eight. Note that, for a slow decay where the coherent parts lead to a homogeneously distributed eight-shape, two interference terms transferred into the ground state at $t/t_{\mathrm{ph}} = (2n+1)/4$ survive the decay process.

The corresponding linear entropies are depicted in Figure 7b, where the blue curves show the linear entropy of the full system S_{lin} and the red ones the linear entropy of the phonons $S_{\mathrm{lin}}^{\mathrm{ph}}$. The decay rate increases from dark to bright colors. We find the same dependency on the decay rate as for a single coherent state in Figure 3b,d, the final entropy increases for a slower decay. In addition, the behavior for very fast decays, e.g., $\Gamma = 5\omega_{\mathrm{ph}}$, is approximately the same as in Figure 3. The entropy of the full system (light blue) forms a sharp peak due to the evolution of the TLS through a statistical mixture, while the phonon part (light red) basically just rises before reaching the stationary value. The dynamics get more involved and new features appear for slow decays, e.g., for $\Gamma = 0.05\omega_{\mathrm{ph}}$. Here, the full entropy in dark blue continuously increases to the stationary value at the end of the decay process, while the phonon contribution in dark red is always slightly smaller and shows additional dynamics developing multiple minima and maxima within each phonon period. While the dynamics are rather irregular on shorter times $t < 3t_{\mathrm{ph}}$, it becomes more periodic for longer periods of time.

To understand the origin of these dynamics in the phonon system, we take a closer look at the different parts of the Wigner function. According to Reference [13], the Wigner function of a cat state can be separated into

$$W = W_{\mathrm{coh}} + W_{\mathrm{int}} \, , \tag{33}$$

where W_{coh} describes the two coherent states that have been studied previously, and W_{int} is the interference showing up as striped structure in phase space. Under the assumption that the phonon coupling strength, γ is large enough such that the different parts of the Wigner function do not significantly overlap in phase space; the linear entropy can also be separated into two contributions, $S_{\mathrm{lin}}^{\mathrm{coh}}$ and $S_{\mathrm{lin}}^{\mathrm{int}}$, calculated from the respective contributions of the Wigner function, and we obtain

$$S_{\mathrm{lin}}^{\mathrm{ph}} \approx S_{\mathrm{lin}}^{\mathrm{coh}} + S_{\mathrm{lin}}^{\mathrm{int}} \, . \tag{34}$$

In Figure 7c, we show the different entropies for a short time window $4 \leqslant t/t_{ph} \leqslant 5$. This already clarifies the picture a bit. First of all, we find that S_{lin}^{coh} (green) and S_{lin}^{int} (blue) are approximately of the same size and the sum of the two parts, shown in dashed blue, agrees perfectly with the full linear phonon entropy (red line). The coherent part S_{lin}^{coh} has reduced values at times $t/t_{ph} = n/2$, and S_{lin}^{int} develops minima exactly between those times, i.e., at $t/t_{ph} = (2n + 1)/4$. Because the shapes of the minima in the two contributions are not the same, the sum appears quite involved.

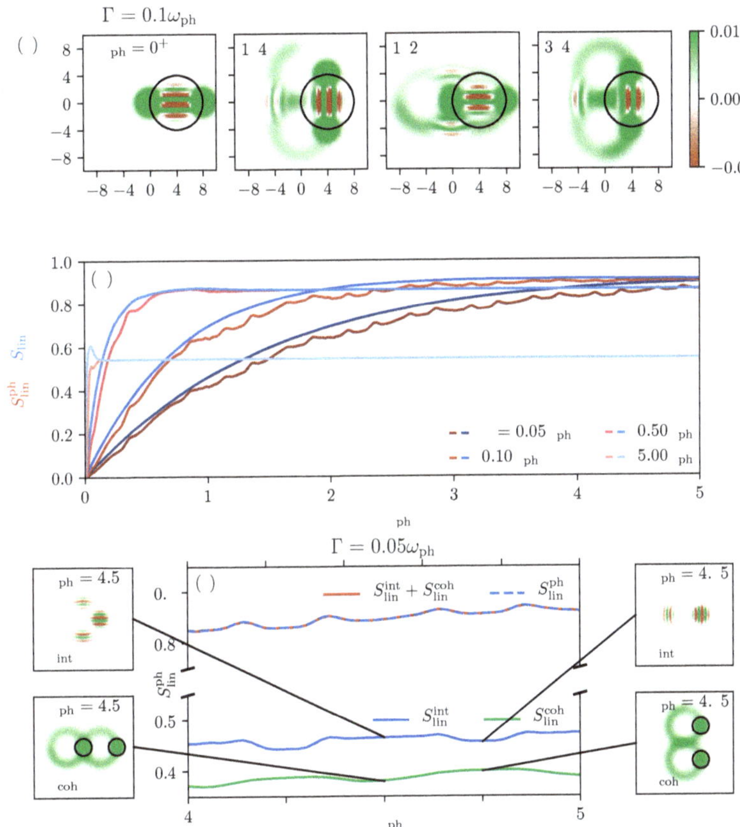

Figure 7. (**a**) Snapshots of the Wigner function during the decay into the ground state for a decay rate of $\Gamma = 0.1\omega_{ph}$. (**b**) Linear phonon entropies in red and full linear entropies in blue as functions of time for different decay rates. Γ increases from bright to dark colors. (**c**) Zoom-in on one phonon period for $\Gamma = 0.05\omega_{ph}$. The coherent contribution S_{lin}^{coh} is green, the one from the interference S_{lin}^{int} is blue and their sum dashed blue. Next to (**c**) are exemplary Wigner functions of the coherent part (bottom) and the interference (top) for the respective minima marked by black lines.

After identifying the different dynamics, we have to understand their origin. Therefore, next to Figure 7c, we plot the Wigner functions W_{coh} and W_{int} for the respective minima, as marked by the black lines. Starting with the coherent part at the bottom, we recognize that the situation is equivalent to the one in Figure 3b,c. The entropy is always reduced when the rotating Wigner function of the coherent states in the excited state subspace (marked by black circles) overlaps with parts of the Wigner function in the ground state subspace. In the right example, at $t = 4.75t_{ph}$, the two Gaussians are clearly separated from the decayed part in the ground state. In the left one, at $t = 4.5t_{ph}$, one of the coherent states overlaps with the touching point of the two circles that are in the ground state $|g\rangle$.

Because we start with two coherent states in $|x\rangle$, the periodicity of the minima is half the phonon period. Moving on to the Wigner function of the interference contribution at the top, we only see the expected striped patterns. The times where $S_{\text{lin}}^{\text{int}}$ is reduced agree with the times $t/t_{\text{ph}} = (2n+1)/4$, where the interference terms that survive the decay process and remain also after the full decay are transferred into the ground state system (see discussion in Reference [32]). As seen in the depicted Wigner function on the right, at these times (e.g., $t = 4.75t_{\text{ph}}$) in Figure 7c, one of the interference terms that were already transferred into $|g\rangle$ perfectly overlaps with the single interference that is still in the excited state, resulting in the two separated structures in phase space. For all other times, three contributions appear, two in $|g\rangle$ and one in $|x\rangle$, as exemplarily shown on the left for $t = 4.5t_{\text{ph}}$. Thus, the fundamental reason for the reduction of the entropy is that a mixture of two cat states is more pure than a mixture of three. The perfectly overlapping interferences on the right make the corresponding Wigner function look more like a mixture of two states than of three.

3.3.1. Phonons Generated by a Decaying TLS

To conclude the discussion, we now take a look at the two pulse excitation discussed in Section 3.2.1 and consider a non-vanishing decay rate of the TLS. In Reference [32], it was shown that the final phonon state is in good agreement with the eight-shaped Wigner function of the decayed single cat state previously analyzed. However, now the phonon generation leads to a statistical mixture of two cat states that are additionally smeared out in phase space. This is exemplarily shown by the Wigner functions in Figure 8a. Although the quantum state of the system is more involved, the linear entropy of the phonons depicted as red line in Figure 8b evolves in a well-structured manner. Especially after the excitation with the second pulse, marked by the dashed black line, the dynamics resemble the ones in Figure 5b with an additional increase of the curve according to the decay process. We find the same broad depressions for half periods and stronger oscillating ones for full periods. The small additional entropy reductions discussed in Figure 7 are also found here, as shown by the zoom-in in Figure 5c. However, compared to the effects of the two overlapping cat states, as previously mentioned, they almost disappear. The linear entropy of the full system shown as a blue line in Figure 5b is always smaller than the phonon part and grows smoothly. This is in agreement with the situation without any decay, where the entropy of the full system was always zero while the phonon part was non-vanishing.

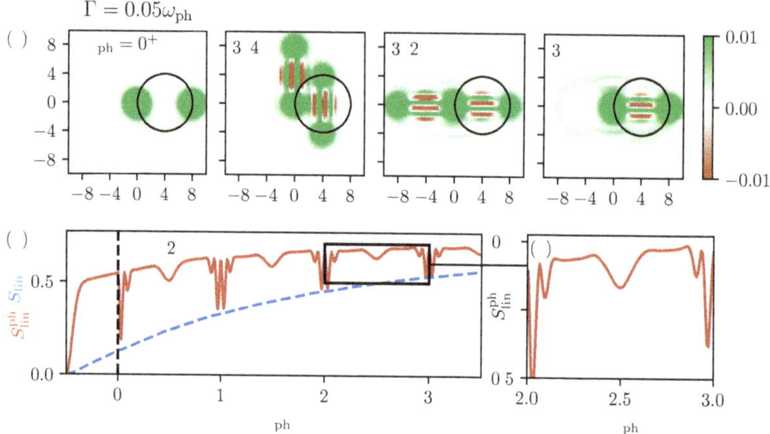

Figure 8. (**a**) Snapshots of the Wigner function's dynamics after the second pulse for $\Gamma = 0.05\omega_{\text{ph}}$. (**b**) Dynamics of the linear phonon entropy in red and the linear entropy of the full system in blue. (**c**) Zoom-in on the marked short time window from (**b**).

4. Conclusions

In summary, we analyzed the entropy dynamics of a single phonon mode coupled to an optically-driven TLS. We presented a theoretical framework that allowed us to calculate entropies of the different parts of the system when the quantum state of the entire system is pure and linear entropies when it is not pure. Additionally, the concept of Wigner functions for the representation of phonon quantum states was used. We started our discussion with the most basic optical excitation, i.e., a single ultrafast pulse, that generated a mixture of two coherent states in the phonon system and assumed a non-decaying TLS. From this, we further increased the complexity of the generated phonon state by including non-vanishing decay rates and two-pulse excitations of the TLS. This led to Wigner functions that smeared out in phase space and the generation of Schrödinger cat states, respectively. While the decay of the TLS, in general, led to an increase of the system's entropy, the complex dynamics of the phonon states resulted in temporally significant reductions of the phonon entropy. All these effects could be traced back to the purity of the quantum states and the entanglement between phonons and TLS. This extensive study on the phonon's entropy led to a thorough understanding of the fundamental interplay between the dynamics of the two separate parts and their combined influence on the quantum state purity.

Author Contributions: conceptualization, T.H., D.W. and T.K.; methodology, T.H.; software, T.H.; validation, T.H., D.W. and T.K.; formal analysis, T.H.; investigation, T.H.; data curation, T.H.; writing–original draft preparation, D.W.; writing–review and editing, T.H., D.W. and T.K.; visualization, T.H. and D.W.; supervision, D.W. and T.K.; project administration, T.K. All authors have read and agreed to the published version of the manuscript.

Funding: This research received no external funding.

Conflicts of Interest: The authors declare no conflict of interest.

References

1. Kondepudi, D.; Prigogine, I. *Modern Thermodynamics: From Heat Engines to Dissipative Structures*; John Wiley & Sons: Berlin, Germany, 2014.
2. Gunzig, E.; Geheniau, J.; Prigogine, I. Entropy and cosmology. *Nature* **1987**, *330*, 621–624. [CrossRef]
3. la Cruz-Dombriz, D.; Sáez-Gómez, D. Black holes, cosmological solutions, future singularities, and their thermodynamical properties in modified gravity theories. *Entropy* **2012**, *14*, 1717–1770. [CrossRef]
4. Toda, M.; Kubo, R.; Saitō, N. *Equilibrium Statistical Mechanics*; Springer: Berlin, Germany, 1992; Volume 1.
5. Kubo, R.; Toda, M.; Hashitsume, N. *Statistical Physics II: Nonequilibrium Statistical Mechanics*; Springer: Berlin, Germany, 2012; Volume 31,
6. Breuer, H.P.; Petruccione, F. *The Theory of Open Quantum Systems*; Oxford University Press: Oxford, UK, 2002.
7. Eisert, J.; Cramer, M.; Plenio, M.B. Colloquium: Area laws for the entanglement entropy. *Rev. Mod. Phys.* **2010**, *82*, 277. [CrossRef]
8. Shannon, C.E. A mathematical theory of communication. *Bell Labs Tech. J.* **1948**, *27*, 379–423. [CrossRef]
9. Liang, J.; Shi, Z.; Li, D.; Wierman, M.J. Information entropy, rough entropy and knowledge granulation in incomplete information systems. *Int. J. Gen. Syst.* **2006**, *35*, 641–654. [CrossRef]
10. Nielsen, M.A.; Chuang, I. *Quantum Computation and Quantum Information*; American Association of Physics Teachers: College Park, MD, USA, 2002.
11. Hofheinz, M.; Wang, H.; Ansmann, M.; Bialczak, R.C.; Lucero, E.; Neeley, M.; O'connell, A.D.; Sank, D.; Wenner, J.; Martinis, J.M.; et al. Synthesizing arbitrary quantum states in a superconducting resonator. *Nature* **2009**, *459*, 546. [CrossRef] [PubMed]
12. O'Connell, A.D.; Hofheinz, M.; Ansmann, M.; Bialczak, R.C.; Lenander, M.; Lucero, E.; Neeley, M.; Sank, D.; Wang, H.; Weides, M.; et al. Quantum ground state and single-phonon control of a mechanical resonator. *Nature* **2010**, *464*, 697–703. [CrossRef] [PubMed]
13. Reiter, D.E.; Wigger, D.; Axt, V.M.; Kuhn, T. Generation and dynamics of phononic cat states after optical excitation of a quantum dot. *Phys. Rev. B* **2011**, *84*, 195327. [CrossRef]

14. Satzinger, K.J.; Zhong, Y.P.; Chang, H.S.; Peairs, G.A.; Bienfait, A.; Chou, M.H.; Cleland, A.Y.; Conner, C.R.; Dumur, É.; Grebel, J.; et al. Quantum control of surface acoustic-wave phonons. *Nature* **2018**, *563*, 661–665. [CrossRef]

15. Mahan, G.D. *Many-Particle Physics*; Plenum Press: New York, NY, USA, 1981.

16. Zrenner, A.; Beham, E.; Stufler, S.; Findeis, F.; Bichler, M.; Abstreiter, G. Coherent properties of a two-level system based on a quantum-dot photodiode. *Nature* **2002**, *418*, 612. [CrossRef]

17. Aharonovich, I.; Castelletto, S.; Simpson, D.A.; Su, C.H.; Greentree, A.D.; Prawer, S. Diamond-based single-photon emitters. *Prog. Phys.* **2011**, *74*, 076501. [CrossRef]

18. Wigger, D.; Karakhanyan, V.; Schneider, C.; Kamp, M.; Höfling, S.; Machnikowski, P.; Kuhn, T.; Kasprzak, J. Acoustic phonon sideband dynamics during polaron formation in a single quantum dot. *Opt. Lett.* **2020**, *45*, 919–922. [CrossRef] [PubMed]

19. Roca, E.; Trallero-Giner, C.; Cardona, M. Polar optical vibrational modes in quantum dots. *Phys. Rev. B* **1994**, *49*, 13704. [CrossRef] [PubMed]

20. Gali, A.; Simon, T.; Lowther, J.E. An ab initio study of local vibration modes of the nitrogen-vacancy center in diamond. *New J. Phys.* **2011**, *13*, 025016. [CrossRef]

21. Debald, S.; Brandes, T.; Kramer, B. Control of dephasing and phonon emission in coupled quantum dots. *Phys. Rev. B* **2002**, *66*, 041301. [CrossRef]

22. Munsch, M.; Kuhlmann, A.V.; Cadeddu, D.; Gérard, J.M.; Claudon, J.; Poggio, M.; Warburton, R.J. Resonant driving of a single photon emitter embedded in a mechanical oscillator. *Nat. Commun.* **2017**, *8*, 76. [CrossRef]

23. Ferry, D.K. *Semiconductors*; Macmillan: New York, NY, USA, 1991.

24. Munn, R.W.; Silbey, R. Theory of exciton transport with quadratic exciton–phonon coupling. *J. Chem. Phys.* **1978**, *68*, 2439–2450. [CrossRef]

25. Muljarov, E.A.; Zimmermann, R. Dephasing in quantum dots: Quadratic coupling to acoustic phonons. *Phys. Rev. Lett.* **2004**, *93*, 237401. [CrossRef]

26. Machnikowski, P. Change of decoherence scenario and appearance of localization due to reservoir anharmonicity. *Phys. Rev. Lett.* **2006**, *96*, 140405. [CrossRef]

27. Chenu, A.; Shiau, S.Y.; Combescot, M. Two-level system coupled to phonons: Full analytical solution. *Phys. Rev. B* **2019**, *99*, 014302. [CrossRef]

28. Duke, C.B.; Mahan, G.D. Phonon-broadened impurity spectra. I. Density of states. *Phys. Rev.* **1965**, *139*, A1965. [CrossRef]

29. Stock, E.; Dachner, M.R.; Warming, T.; Schliwa, A.; Lochmann, A.; Hoffmann, A.; Toropov, A.I.; Bakarov, A.K.; Derebezov, I.A.; Richter, M.; et al. Acoustic and optical phonon scattering in a single In (Ga) As quantum dot. *Phys. Rev. B* **2011**, *83*, 041304. [CrossRef]

30. Wigger, D.; Schneider, C.; Gerhardt, S.; Kamp, M.; Höfling, S.; Kuhn, T.; Kasprzak, J. Rabi oscillations of a quantum dot exciton coupled to acoustic phonons: Coherence and population readout. *Optica* **2018**, *5*, 1442–1450. [CrossRef]

31. Ramsay, A.J.; Godden, T.M.; Boyle, S.J.; Gauger, E.M.; Nazir, A.; Lovett, B.W.; Fox, A.M.; Skolnick, M.S. Phonon-induced Rabi-frequency renormalization of optically driven single InGaAs/GaAs quantum dots. *Phys. Rev. Lett.* **2010**, *105*, 177402. [CrossRef]

32. Hahn, T.; Groll, D.; Kuhn, T.; Wigger, D. Influence of excited state decay and dephasing on phonon quantum state preparation. *Phys. Rev. B* **2019**, *100*, 024306. [CrossRef]

33. Auffeves, A.; Richard, M. Optical driving of macroscopic mechanical motion by a single two-level system. *Phys. Rev. A* **2014**, *90*, 023818. [CrossRef]

34. Schleich, W.P. *Quantum Optics in Phase Space*; John Wiley & Sons: Berlin, Germany, 2011.

35. Gerry, C.; Knight, P.L. *Introductory Quantum Optics*; Cambridge University Press: Cambridge, UK, 2005.

36. Von Neumann, J. *Mathematische Grundlagen der Quantenmechanik*; Springer: Berlin, Germany, 1996; Volume 2.

37. Wehrl, A. General properties of entropy. *Rev. Mod. Phys.* **1978**, *50*, 221. [CrossRef]

38. Manfredi, G.; Feix, M.R. Entropy and Wigner functions. *Phys. Rev. E* **2000**, *62*, 4665. [CrossRef]

39. Brune, M.; Haroche, S.; Raimond, J.M.; Davidovich, L.; Zagury, N. Manipulation of photons in a cavity by dispersive atom-field coupling: Quantum-nondemolition measurements and generation of "Schrödinger cat"states. *Phys. Rev. A* **1992**, *45*, 5193. [CrossRef]

40. Ourjoumtsev, A.; Jeong, H.; Tualle-Brouri, R.; Grangier, P. Generation of optical 'Schrödinger cats' from photon number states. *Nature* **2007**, *448*, 784–786. [CrossRef]
41. Deleglise, S.; Dotsenko, I.; Sayrin, C.; Bernu, J.; Brune, M.; Raimond, J.M.; Haroche, S. Reconstruction of non-classical cavity field states with snapshots of their decoherence. *Nature* **2008**, *455*, 510. [CrossRef] [PubMed]

MDPI

St. Alban-Anlage 66

4052 Basel

Switzerland

Tel. +41 61 683 77 34

Fax +41 61 302 89 18

www.mdpi.com

Entropy Editorial Office

E-mail: entropy@mdpi.com

www.mdpi.com/journal/entropy